CAM 与数控编程（英文版）

CAM and Numerical Control Programming

主　编　封志明

主　审　周利平

西南交通大学出版社
·成　都·

模型源文件下载

内 容 简 介

本英文教材是结合近年来国外出版的英文原版专著、教材,并参考 Siemens NX 软件技术说明文编写而成。本书主要介绍数控加工编程、CAM 基础知识、CAM 操作方法及工作流程。在介绍 CAM 时,以 UG NX7.5 为平台,系统介绍了常用的加工操作方式及其加工创建、参数设置、机床控制、实例仿真和后置处理。为了提高本书的实用性,采用项目教学法针对每个应用模块给出了相应的典型操作实例,同时在最后一章还给出了大型综合实例的自动编程指导。为了便于读者学习,在本书的每一章都对一些疑难句做了详细注释并给出了专业词汇表,全书最后给出了总词汇表。本书还为授课教师准备了所有案例的源文件。

该教材可作为机械制造及其自动化、机械电子工程等专业的高年级本科生的双语教材,也可用作其他有关专业人员 CAM 数控编程的英文培训教材。

图书在版编目(CIP)数据

CAM 与数控编程 = CAM and Numerical Control Programming:英文 / 封志明主编. —成都:西南交通大学出版社,2016.8
 ISBN 978-7-5643-4843-4

Ⅰ. ①C… Ⅱ. ①封… Ⅲ. ①计算机辅助制造 – 英文 ②数控机床 – 程序设计 – 英文 Ⅳ. ①TP391.73②TG659

中国版本图书馆 CIP 数据核字(2016)第 175641 号

CAM 与数控编程(英文版)

CAM and Numerical Control Programming

主编 封志明

责 任 编 辑	宋彦博
助 理 编 辑	张文越
封 面 设 计	何东琳设计工作室
出 版 发 行	西南交通大学出版社 四川省成都市二环路北一段 111 号 西南交通大学创新大厦 21 楼)
发 行 部 电 话	028-87600564　028-87600533
邮 政 编 码	610031
网　　　　址	http://www.xnjdcbs.com
印　　　　刷	四川煤田地质制图印刷厂
成 品 尺 寸	185 mm × 260 mm
印　　　　张	16.75
字　　　　数	544 千
版　　　　次	2016 年 8 月第 1 版
印　　　　次	2016 年 8 月第 1 次
书　　　　号	ISBN 978-7-5643-4843-4
定　　　　价	39.80 元

图书如有印装质量问题　本社负责退换
版权所有　盗版必究　举报电话:028-87600562

前 言

随着计算机技术的发展，计算机辅助设计与制造（CAD/CAM）被越来越广泛地应用于航空航天、汽车、模具及精密机械等各个领域。随着全球制造业中心向我国逐步转移，国内对数控编程人员的需求呈现高速、持续的增长趋势。

UG NX 是目前全球领先的产品生命周期管理和制造运营管理软件提供商，提供从规划、开发，直至制造、生产和技术支持在内的生命周期管理功能。NX CAM 提供了完整的高速曲面加工、多功能铣削-车削加工以及五轴加工功能。通过使用 NX CAM，NC 编程人员能够使其最新、最高效和功能最强的机床产生最大的价值。

同先进的工业国家相比，我国制造业技术水平还很低。为了使工科高年级学生能够尽快熟悉本专业的技术词汇，了解国外先进制造技术发展及软件应用情况，我们参考了近年来国外出版的关于数控技术的英文原版教材，以及介绍 CAM 数控编程的相关文章和报告，并结合 Siemens NX 官方网站关于 CAM 应用模块的技术说明文档资料，编写了该双语教材。我们鼓励学生使用本专业领域最新的外文原版教材，阅读本专业外文原版学习资料，以便在知识更新越来越快的信息时代能够及时了解最新科研成果，并使用在装备制造领域最先进的外文原版软件。

本书重点介绍了 NX7.5 CAM 模块的基本功能，辅以大量的图形进行讲解，让读者一目了然，同时配合精选的编程案例，实现基于项目的学习，让读者在学习过程中可以亲手创建高效、高质的刀位轨迹，以便快速掌握数控编程的基本知识和技能。本书通过项目教学法，力求培养读者全面完整的设计思想，达到融会贯通、举一反三的目的，使其成为合格的 CAM 数控编程工程师。

为了提高本教材的教学效果，我们特意在每章后面附有一些疑难句子的注释及该章节的词汇表，并在书后附有总词汇表。同时，为了帮助读者更加直观地学习，提供了全书各实例的源文件。本书可作为机械制造及其自动化、机械电子工程等专业的高年级本科生的双语教材，也可用作其他有关专业人员 CAM 数控编程的英文培训教材。

本书由西华大学机械工程学院封志明博士任主编，由周利平教授任主审，西华大学宋敏莉完成了第1章，第2章的编写，并提供了本书中的所有案例源文件。在编写过程中，我们得到了西华大学外国语学院彭旭老师的大力支持与帮助，在此表示衷心的感谢。

限于作者的知识水平，本书在编写过程中难免有疏漏和不足之处，恳请广大读者和同仁不吝赐教，对书中不足之处给予指正。

编　者

2016 年 4 月

Contents

Chapter 1 Fundamentals of CNC Machining ... 1
1.1 Numerical Control Fundamentals ... 1
1.2 CNC part programming ... 8
1.3 Computer aided manufacturing (CAM) ... 12

Chapter 2 Basis of UG NX CAM ... 18
2.1 UG NX CAM overview ... 18
2.2 UG CAM machining environment ... 22
2.3 Analyzing the part to machine ... 24
2.4 Creating CAM operations ... 26
2.5 Coordinate system of UG NX CAM ... 35
2.6 Tool Paths ... 37
2.7 Postprocessing ... 41
2.8 Operation Navigator ... 46

Chapter 3 UG CAM Path Settings ... 52
3.1 Path settings overview ... 52
3.2 Cut Pattern ... 53
3.3 Stepover ... 54
3.4 Cut levels ... 55
3.5 Cutting Parameters ... 55
3.6 Non-cutting Moves ... 70
3.7 Feeds and speeds ... 79
3.8 Machine control ... 81

Chapter 4 Planar Milling ... 85
4.1 Planar milling & Path Settings overview ... 85
4.2 Manufacturing boundaries overview ... 86
4.3 Planar milling geometry ... 92
4.4 Planar milling operation parameters ... 94
4.5 Planar milling cutting parameters ... 101
4.6 Create planar milling operation ... 104
4.7 Planar milling example ... 105

Chapter 5 Cavity Milling ... 114
5.1 Cavity milling overview ... 114
5.2 In Process Workpiece (IPW) ... 115
5.3 Setting cavity milling geometry ... 121

5.4	Cavity milling cutting parameters	124
5.5	Create cavity milling operation	132
5.6	Cavity Milling Operation Example	133

Chapter 6 Face Milling · 141

6.1	Face milling overview	141
6.2	Face milling geometry	142
6.3	Face milling operation parameters	145
6.4	Face milling cutting parameters	147
6.5	Create face milling operation	151
6.6	Face milling example	152

Chapter 7 Z-Level Milling · 157

7.1	Z-level milling overview	157
7.2	Z-level milling geometry	158
7.3	Z-level milling operation parameters	159
7.4	Z-level milling cutting parameters	160
7.5	Z-level milling example	165

Chapter 8 Fixed-axis Surface Contouring · 171

8.1	Surface contouring overview	171
8.2	Valid geometry for fixed-axis surface contouring	173
8.3	Drive methods	173
8.4	Projection Vector	193
8.5	Tool axis	194
8.6	Create a fixed contouring operation	194
8.7	Fixed contouring operation example	195

Chapter 9 Drilling · 201

9.1	Drilling operations overview	201
9.2	Drill geometry	204
9.3	Drilling cycles	211
9.4	Cycle Parameters	216
9.5	Drilling operation example	220

Chapter 10 A Comprehensive CAM Instance · 227

10.1	Part analysis	227
10.2	Setting up the machining environment	229
10.3	Preparation for creating operations	229
10.4	Create milling operations	233

Vocabulary List · 256

Reference · 261

Chapter 1
Fundamentals of CNC Machining

Objectives:
- ✓ To understand the working principle and applications of CNC machines.
- ✓ To be able to prepare CNC part programs for machining 2-D workpieces.
- ✓ To understand the structure and flow of a CAM system.

1.1　Numerical Control Fundamentals

1.1.1　CNC System Overview

Computer Numerical Control (CNC) is a specialized and versatile form of flexible automation and is widely applied in many areas, although it was initially developed to control the motion and operation of machine tools.

CNC may be considered to be a means of operating a machine through the use of discrete numerical values fed into the machine, where the required input technical information is stored on a kind of input media such as floppy disk, CD ROM, USB flash drive, or RAM card. etc.. The machine follows a predetermined sequence of machining operations at the predetermined speeds, which are necessary to produce a workpiece of the right shape and size, according to completely predictable results. A different product can be produced through reprogramming and a low-quantity production run of different products is justified. Fig.1-1 shows the CNC machine centre.

Fig.1-1　CNC machine centre

The definition of CNC given by Electronic Industry Association (EIA) is as follows:

A system in which actions are controlled by the direct insertion of numerical data at some point. The system must automatically interpret at least some portion of this data.

In a simple word, a CNC system receives numerical data, interprets the data and then controls the action accordingly.

1.1.2 Control Systems

A CNC machine can be controlled through two type of circuits: open-loop and closed-loop. The overall precision of the machine is determined by the type of control loop used.

1. Open-loop systems

Open-loop systems (Fig.1-2) have no access to the real time data about the performance of the system and therefore no immediate corrective action can be taken in case of system disturbance. This system is normally applied only to the case where the output is almost constant and predictable. Therefore, an open-loop system is unlikely to be used to control machine tools since the cutting force and loading of a machine tool is never a constant. The only exception is the wire-cut machine for which some machine tool builders still prefer to use an open-loop system because there is virtually no cutting force in wire-cut machining.

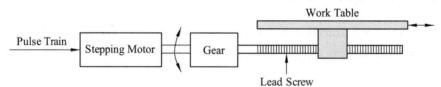

Fig.1-2 Block Diagram of an Open-loop System

2. Closed-loop systems

In a closed-loop system (Fig.1-3), feed back devices closely monitor the output and any disturbance will be corrected in the first instance. Therefore high system accuracy is achievable. This system is more powerful than the open-loop system and can be applied to the case where the output is subjected to frequent change. Closed-loop systems are very accurate. Most of them have auto compensation for error, since the feedback devices indicate the error and then control makes the necessary adjustments to bring the slide back to the position. They used AC, DC, or hydraulic servomotors. Nowadays, almost all CNC machines use this control system.

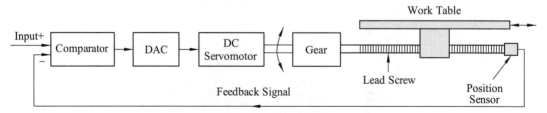

Fig.1-3 Block Diagram of a Closed-loop System

1.1.3 Construction of a CNC system

In general, a typical CNC system consists of the following 6 major elements.
- Input Device.
- Machine Control Unit (MCU).
- Machine Tool.
- Driving System.
- Feedback Devices.
- Display Unit.

The working principles of CNC machines are shown in Fig.1-4.

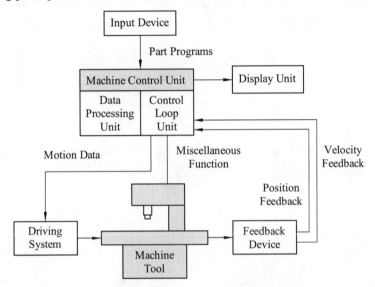

Fig.1-4 Working principles of CNC machines

1. Input devices

Input devices store the required input technical information. The commonly used input media include several parts.

(1) USB flash drive.

A USB flash drive is a removable, rewritable and portable hard drive with compact size and bigger storage size than a floppy disk. Data stored inside the flash drive. CNC are impervious to dust and scratches that enable flash drives to transfer data from place to place. In recent years, all computers support USB flash drives to read and write data, which make it become more and more popular in CNC machine control unit.

(2) Serial communication.

The data transferring between a computer and a CNC machine tool is often accomplished through a serial communication port. International standards for serial communications are established so that information can be exchanged in an orderly way. The most common interface between computers and CNC machine tools is referred to the EIA Standard RS-232. Most of the personal computers and CNC machine tools have built in RS-232 port and a standard RS-232 cable is used to connect a CNC machine to a computer which enables the data transfer in reliable way. Part programs can be downloaded into the memory of a machine tool or uploaded to the computer for temporary storage by running a communication program on the computer and setting up the machine control to interact with the communication software.

Distributed numerical control is a hierarchical system for distributing data between a production management computer and NC systems. The host computer is linked with a number of CNC machines or computers connecting to the CNC machines for downloading part programs. The communication program in the host computer can utilize two-way data transfer features for production data communication, including: production schedule, parts produced and machine utilization, etc. Serial communication in distributed numeric control system is indicated in Fig.1-5.

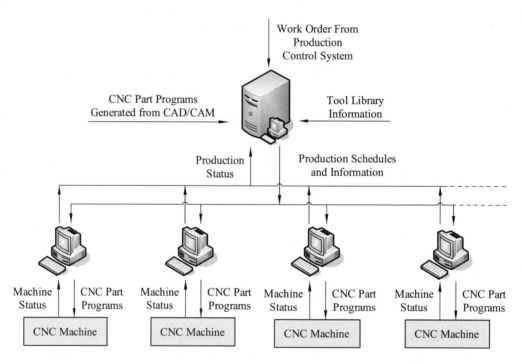

Fig.1-5 Serial communication in a distributed numerical control system

(3) Ethernet communication.

Due to the advancement of the computer technology and the drastic reduction of the cost of the computer, it is becoming more practical and economic to transfer part programs between computers and CNC machines via an Ethernet communication cable. This medium provides a more efficient and reliable means in part programs transmission and storage. Most companies now built a Local Area Network (LAN) as their infrastructure. More and more CNC machine tools provide an option of the Ethernet card for direct communication within the LAN. Ethernet network in a distributed numerical control system is indicated in Fig.1-6.

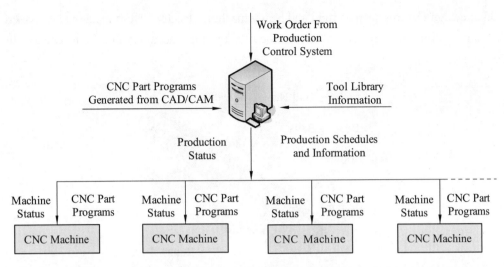

Fig.1-6 Ethernet network in a distributed numerical control system

(4) Conversational programming.

Part programs can be input to the controller via the keyboard. Built-in intelligent software inside the controller enables the operator to enter the required data step by step. This is a very efficient way for preparing programs for relatively simple workpieces involving up to 2.5 axis machining.

2. Machine Control Unit (MCU)

The machine control unit is the heart of the CNC system. There are two sub-units in the machine control unit: the Data Processing Unit (DPU) and the Control Loop Unit (CLU).

(1) Data Processing Unit (DPU).

On receiving a part program, the DPU firstly interprets and encodes the part program into internal machine codes. The interpolator of the DPU then calculate the intermediate positions of the motion in terms of BLU (Basic Length Unit) which is the smallest unit length that can be handled by the controller. The calculated data are passed to CLU for further action.

(2) Control Loop Unit (CLU).

The data from the DPU are converted into electrical signals in the CLU to control the driving system to perform the required motions. Other functions such as machine spindle on/off, coolant on/off, tool clamp on/off are also controlled by this unit according to the internal machine codes.

3. Machine tool

This can be any type of machine tool or equipment. In order to obtain high accuracy and repeatability, the design and manufactrue of the machine slide and the driving lead screw of a CNC machine is of vital importance. The slides are usually machined to high accuracy and coated with anti-friction material in order to reduce the stick and slip phenomenon. Large diameter recirculating ball screws are employed to eliminate the backlash and lost motion. Fig1-7 shows the ball screw structure.

Other design features such as rigid and heavy machine structure, short machine table overhang, quick change tooling system, etc. also contribute to the high accuracy and high repeatability of CNC machines.

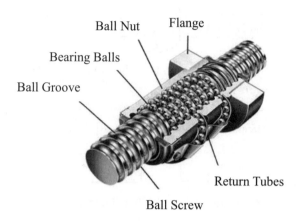

Fig.1-7 Ball Screw Structure

4. Driving system

The driving system is an important component of a CNC machine as the accuracy and repeatability depending very much on the characteristics and performance of the driving system. The requirement is that the driving system has to response accurately according to the programmed instructions. This system usually uses electric motors although hydraulic motors are sometimes used for large machine tools. The motor is coupled either directly or through a gear box to the machine, which leads screw to move the machine slide or the spindle. Four types of electrical motors are commonly used.

(1) DC servomotor.

Direct Current (DC) servomotor is the most common type of feed motors used in CNC machines. The principle of operation is based on the rotation of an armature winding in a permanently energized magnetic field. They are used to drive a lead screw and gear mechanism.

(2) AC servomotor.

In an Alternating Current (AC) servomotor, the rotor is a permanent magnet while the stator is equipped with 3-phase windings. The speed of the rotor is equal to the rotational frequency of the magnetic field of the stator, which is regulated by the frequency converter. AC motors are gradually replacing DC servomotors.

(3) Stepper motor.

A stepper motor is a device that converts the electrical pulses into discrete mechanical rotational motions of the motor shaft. This is the simplest device that can be applied to CNC machines since it can convert digital data into actual mechanical displacement. It is not necessary to have any analog-to-digital converter nor feedback device for the control system. They are ideally suited to open-loop systems. Stepper motors are mostly used in applications where low torque is required.

(4) Fluid servomotor.

Fluid servomotors are also variable-speed motors. They are able to produce more power, or more speeds in the case of pneumatic motors than electric servomotors.

5. Feedback device

In order to have a CNC machine operating accurately, the positional values and speed of the axes need to be constantly updated. Two types of feedback devices are normally used: positional feedback device and velocity feedback device.

(1) Positional feedback devices.

There are two types of positional feedback devices: linear transducer for direct positional measurement and rotary encoder for angular or indirect linear measurement.

A linear transducer is a device mounted on the machine table to measure the actual displacement of the slide in such way that backlash of screws and motors would not cause any error in the feedback data. This device is considered to be of the highest accuracy and also more expensive in comparison with other measuring devices mounted on screws or motors.

A rotary encoder is a device mounted at the end of the motor shaft or screw to measure the angular displacement. This device cannot measure linear displacement directly so that errors may occur due to the backlash of screws and motors. Generally, this error can be compensated by the machine builder in the machine calibration process.

(2) Velocity feedback device.

The actual speed of the motor can be measured in terms of voltage generated from a tachometer mounted at the end of the motor shaft. DC tachometer is essentially a small generator that produces an output voltage proportional to the speed. The voltage generated is compared with the command voltage corresponding to the desired speed. The difference of the voltages is then used to actuate the motor to eliminate the error.

6. Display unit

The display unit (see Fig.1-8) serves as an interactive device between the machine and the operator. When the machine is running, the display unit displays the present CNC status such as the position of the machine slide, the spindle RPM, the feed rate, the part programs, etc.. In an advanced CNC machine, the display unit can show the graphics simulation of the tool path so that part programs can be verified before the actually machining. Much other important information about the CNC system can also display for maintenance and installation work such as machine parameters, logic diagram of the programmer controller, error massages and diagnostic data.

Fig.1-8   Display unit for CNC machines

1.1.4 Applications of CNC machines

CNC machines are widely used in the metal cutting industry and are best used to produce the following types of products.
- Parts with complicated contours.
- Parts requiring close tolerance and/or good repeatability.
- Parts requiring expensive jigs and fixtures if they are produced on conventional machines.
- Parts that may have several engineering changes, such as during the development stage of a prototype.
- In cases where human errors could be extremely costly.
- Parts that are needed in a hurry.
- Small batch lots or short production runs.

Some common types of CNC machines and instruments used in industry are as following.
- Drilling machine.
- Lathe / Turning centre.
- Milling / Machining centre.
- Grinding machine.
- Electro discharge machine (EDM).
- Laser cutting machine.
- Water jet cutting machine.

1.2 CNC part programming

1.2.1 Axis of motion

In generally, all motions have 6 degrees of freedom. In other words, motion can be resolved into 6 axes, namely, 3 linear axes (X, Y and Z axis) and 3 rotational axes (A, B, and C axis). Fig.1-9 shows the axis of motion.

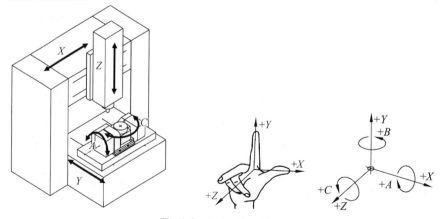

Fig.1-9 Axis of motion

1.2.2 Dimension systems

1. Incremental system

This type of control is always used as a reference to the preceding point in a sequence of points. The disadvantage of this system is that if an error occurs, it will be accumulated. Fig.1-10 (a) shows the incremental system.

2. Absolute system

In an absolute system, all references are made to the origin of the coordinate system. All commands of motion are defined by the absolute coordinate referred to the origin. Fig.1-10 (b) shows the absolute system.

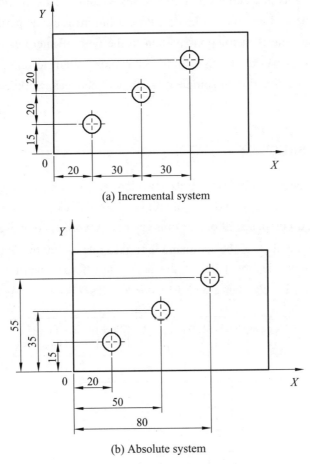

(a) Incremental system

(b) Absolute system

Fig.1-10 Dimension systems

1.2.3 Definition of programming

NC programming is where all the machining data are compiled and where the data are translated into a language which can be understood by the control system of the machine tool. The machining data is as follows.

① Machining sequence: classification of process, tool start up point, cutting depth, tool path, etc..

② Cutting conditions: spindle speed, feed rate, coolant, etc..

③ Selection of cutting tools.

A program for numerical control consists of a sequence of directions that causes an NC machine to carry out a certain operation, while machining is the most commonly used process. Programming for NC may be done by an internal programming department, on the shop floor, or purchased from an outside source. Also, programming can be done manually or with computer assistance.

The program contains instructions and commands. Geometric instructions pertain to relative movements between the tool and the workpiece. Processing instructions pertain to spindle speeds, feeds, tools, and so on. Travel instructions pertain to the type of interpolation and slow or rapid movements of the tool or worktable. Switching commands pertain to on/off position for coolant supplies, spindle rotation, direction of spindle rotation, tool changes, workpiece feeding, clamping, and so on.

1.2.4 Program structure

A CNC program consists of blocks, words and addresses, as shown in Fig.1-11.

· Block: A command given to the control unit is called a block.

· Word: A block is composed of one or more words. A word is composed of an identification letter and a series of numerals, e.g. the command for a feed rate of 200 mm/min is F200.

· Address: The identification letter at the beginning of each word is called address. The meaning of the address is in accordance with EIA standard RS-274-D.

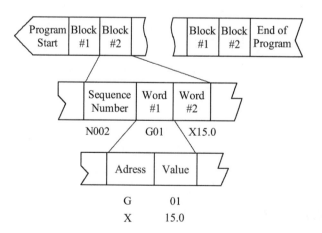

Fig.1-11 Structure of CNC part program

The commonly used function address of CNC program is shown in Tab.1-1.

Tab.1-1 Commonly used function address

Function	Address
Sequence number	N
Preparatory function	G
Coordinate word	X,Y,Z
Parameters for circular interpolation	I,J,K
Feed function	F
Spindle function	S
Tool function	T
Miscellaneous function	M

An example of a program is as follows:

N10 G01 X20.5 F200 S1000 M03
N11 G02 X30.0 Y40.0 I20.5 J32.0

1.2.5 Steps for CNC programming and machining

The following is the procedures to be followed in CNC programming and machining. The most important point is to verify the program by test run it on the machine before the actual machining in order to ensure that the program is free of mistakes.

· Study the part drawing carefully.
· Unless the drawing dimensions are CNC adapted, select a suitable program zero point on the work piece. The tool will be adjusted to this zero point during the machine setup.
· Determine the machining operations and their sequence.
· Determine the method of work clamping (vice, rotary table, fixtures, etc.).
· Select cutting tools and determine spindle speeds and feeds.
· Write program (translate machining steps into program blocks). If many solutions are possible, try the simplest solution first. It is usually longer, but better to proceed in this way.
· Prepare tool chart or diagram, measure tool geometry (lengths, radii) and note.
· Clamp work piece and set up machine.
· Enter compensation value if necessary.
· Check and test program. It is a good practice to dry run the program without the workpiece, without the cutting tools, or by raising the tool to a safe height. If necessary, correct and edit program and check again.
· Start machining.

1.3 Computer aided manufacturing (CAM)

1.3.1 Computer aided part programming

In manual preparation of a CNC part program, the programmer is required to define the machine or the tool movement in numerical terms. If the geometry is complicated 3D surfaces, it cannot be programmed manually.

Over the past years, lots of efforts are devoted to automate the part program generation. With the development of the CAD/CAM system, interactive graphic system is integrated with the CNC part programming. Graphic-based software using menu driven technique improves the user friendliness. The part programmer can create the geometrical model in the CAM package or directly extract the geometrical model from the CAD/CAM data base. Tool motion commands can assist the part programmer to calculate the tool paths automatically. The part programmer can verify the tool paths through the graphic display using the animation function of the CAM system. It greatly enhances the speed and accuracy in tool path generation.

1.3.2 Flow of A Computer Aided Manufacturing System

There are several computer aided manufacturing or CAD/CAM system available in the market. The flow chart of a CAM system is shown in Fig.1-12. Their basic features can be summarized below:

① Geometric modeling / CAD interface.
② Tool motion definition.
③ Data processing.
④ Post processing.
⑤ Data transmission.

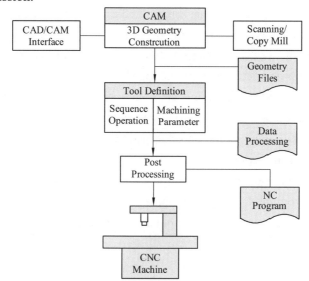

Fig.1-12 Flow Chart of a CAM System

1. Geometric modeling / CAD interface

The geometry of the workpiece can be defined by basic geometrical elements such as points, lines, arcs, splines or surfaces. The two dimensional or three dimensional geometrical elements are stored in the computer memory in forms of a mathematical model. The mathematical model can be a wire frame model, a surface model, or a solid model.

In addition, the geometric models can be imported from other CAD/CAM system through standard CAD/CAM interface formats such as Initial Graphic Exchange Specification (IGES). IGES is a graphic exchange standard jointly developed by industry and the National Bureau of Standards with the support of the U.S. Air Force. It provides transportability of 3-dimensional geometry data between different systems.

Throughout this system, geometrical elements from one system can be translated into a neutral file standard and then from this standard into other format.

2. Tool motion definition

After the geometric modeling, machining data such as the job setup, operation setup and motion definition are input into the computer to produce the cutting location file for machining the workpiece.

(1) Job setup.

This is used to input the machine datum, home position, and the cutter diameters for the cutting location file.

(2) Operation setup.

This is applied to input the system the operation parameters such as the feed rate, tolerance, and approach / retract planes, spindle speed, coolant on/off, stock offset and the tool selection, etc..

(3) Motion definition.

Built-in machining commands are used to control the tool motion to machine the products. This includes the hole processing, profile machining, pocketing, surface machining, gouge checking, etc.

3. Data Processing

The input data is translated into computer usable format. The computer will process the desired part surface, the cutter offset surface and finally compute the paths of the cutter which is known as the cutter location data file. The tool paths can normally be animated graphically on the display for verification purpose.

Furthermore, production planning data such as tool list, setup sheet, and machining time is also calculated for users' reference.

4. Post processing

Different CNC machines have different features and capabilities, the format of the CNC program may also vary from each other. A process is required to change the general instructions from the cutter location file to a specific format for a particular machine tool and this process is called postprocessing.

Post processor is a computer software which converts the cutter location data files into a format which the machine controller can interpret correctly. Generally, there are two types of postprocessor.

(1) Specific postprocessor.

This is a tailor-made software which output the precise code for a specific CNC machine. The user is not required to change anything in the program.

(2) Generic (Universal) postprocessor.

This is a set of generalized rules which needs the user to customize into the format that satisfies the requirements of a specific CNC machine.

5. Data Transmission

After postprocessing, the CNC program can be transmitted to the CNC machines either through the off line or on line process.

(1) Off-line Processes.

Data carriers are used to transmit the CNC program to the CNC machines. It includes paper tapes, magnetic tape or magnetic disc.

(2) On-line Processes.

On-line processes is commonly used in DNC operation and data is transferred either serially or parallel using data cables.

1.3.3 The commonly used CAM Software

1. Unigraphics (UG) NX

Siemens NX software is an integrated product design, engineering and manufacturing solution that helps you deliver better products faster and more efficiently.

NX for Manufacturing provides you with a complete solution set for part manufacturing – from CAM to CNC controller, including machine tool programming, postprocessing and machining simulation, etc.. By using NX for manufacturing, you can improve your part manufacturing productivity with ways listing below:

· Reduce NC programming and machining time.

· Produce better quality parts.

· Maximize use of manufacturing resources.

· maximize returns on your investments in the latest machine tool technology.

Adopted across many industries, NX CAM software delivers proven capabilities for manufacturing in the aerospace, automotive, medical device, mold and die, and machinery industries. The flexibility of NX CAM software means that you can easily complete the most demanding jobs.

Since NX is the industry's most integrated, flexible and efficient solution for product design, engineering and manufacturing, the book introduces the CAM solution based on Siemens NX 7.5.

2. Mastercam

Mastercam is the most widely used CAM software worldwide and remains the program of choice among CNC programmers. Mastercam mill delivers the most comprehensive milling package with powerful new tool paths and techniques, giving the best possible foundation for fast and efficient milling. From general purpose methods such as optimized pocketing to highly specialized tool paths like 5-axis turbine cutting, Mastercam ensures that you're ready for any job. Mastercam delivers:

· Full 3D CAD modeling.
· Easy pocketing, contouring and drilling.
· Feature-based 2D programming.
· Robust 3D solids and surface machining.
· Powerful multi-axis cutting.
· Specialized options for focused projects.

3. Cimatron

Cimatron is part of 3D Systems, a leading provider of 3D printing centric design-to-manufacturing solutions including 3D printers, print materials and cloud sourced on-demand custom parts for professionals and consumers alike in materials including plastics, metals, ceramics and edibles.

GibbsCAM is one of the major CAD/CAM softwares for manufacturing product lines. It is a powerful and easy-to-use CAM software for CNC machine programming that provides seamless integration for CAD package. GibbsCAM's wide range of programming capabilities provide solutions for 2 to 5-axis milling, turning, multi-task machining, swiss-style machining, and wire EDM. With GibbsCAM, CNC programming is flexible, fast, reliable and efficient.

In addition, there are also some other companies which provide CAM software, such as CATIA, PRO/E, CAXA, etc..

 Notes

[1] CNC be considered to be a means of operating a machine through the use of discrete numerical values fed into the machine, where the required input technical information is stored on a kind of input media such as floppy disk, CD ROM, USB flash drive, or RAM card, etc..

计算机数字控制是一种通过向机器输入离散数值从而实现机器运行的一种手段。其中，所需输入的技术信息通常储存在一些输入装置中，如软盘、光驱、优盘或随机存取内存卡等。

[2] Due to the advancement of the computer technology and the drastic reduction of the cost of the computer, it is becoming more practical and economic to transfer part programs between computers and CNC machines via an Ethernet communication cable.

由于计算机技术的进步和成本的大幅降低，在计算机和数控机床之间通过以太网通信电缆传输部件加工程序变得更加可行和经济。

[3] A linear transducer is a device mounted on the machine table to measure the actual displacement of the slide in such way that backlash of screws and motors would not cause any error in the feedback data.

线性传感器安装在机床工作台上，用来测量实际的滑动位移，通过这种方式使得螺钉和电动机的反冲不会造成任何反馈数据错误。

[4] NC programming is where all the machining data are compiled and where the data are translated into a language which can be understood by the control system of the machine tool.

数控编程使得所有加工数据都得以编译，并且这些数据都转换成了一种能被机床控制系统理解的语言。

[5] A process is required to change the general instructions from the cutter location file to a specific format for a particular machine tool and this process is called postprocessing.

为一个特定的机床将刀具位置文件从通用的指令转换为特定的格式需要一个过程，该过程称为后置处理。

New Words in Chapter 1

flexible automation			柔性自动化
Electronic Industry Association(EIA)			电子工业协会
open-loop		adj.	开环的
closed-loop	['kləʊzdluːp]	adj.	闭合环路的，闭环的
precision	[prɪ'sɪʒ(ə)n]	n.	精度，精确
constant	['kɒnst(ə)nt]	n.	常数；恒量
wire-cut	['waɪəkʌt]	adj.	线切割
accuracy	['ækjʊrəsɪ]	n.	精确度，准确性
feedback	['fiːdbæk]	n.	反馈
hydraulic	[haɪ'drɒlɪk]	adj.	液压的，水力的
servomotor	['sɜːvəʊˌməʊtə]	n.	伺服电动机，继动器
serial communication			串行通信
cable	['keɪb(ə)l]	n.	电缆
hierarchical	[haɪə'rɑːkɪk(ə)l]	adj.	分层的
Distributed Numerical Control(DNC)			分布式数字控制
Ethernet	['iːθənet]	n.	以太网
Machine Control Unit (MCU)			机床控制器
Data Processing Unit (DPU)			数据处理器
Control Loop Unit (CLU)			环路控制器

interpolator	[ɪn'tə:pəuleɪtə]		插补器
Basic Length Unit(BLU)			基本长度单元
spindle	['spɪnd(ə)l]	n.	主轴
coolant	['kuːl(ə)nt]	n.	冷却剂
clamp	[klæmp]	n.	夹钳，螺丝钳
screw	[skruː]	n.	螺旋，螺丝钉
backlash	['bæklæʃ]	n.	反冲
overhang	[əʊvə'hæŋ]	n.	悬垂
gear	[gɪə]	n.	齿轮，传动装置
Direct Current (DC)			直流电
armature	['ɑːmətʃə; -tjʊ)ə]	n.	电枢
magnetic	[mæg'netɪk]	adj.	地磁的，有磁性的
mechanism	['mek(ə)nɪz(ə)m]	n.	机械装置
Alternating Current(AC)			交流电
stator	['steɪtə]	n.	固定片，定子
stepper motor			步进电机
torque	[tɔːk]	n.	转矩，[力] 扭矩
fluid	['fluːɪd]	n.	流体，液体
transducer	[trænz'dusə]	n.	传感器，变换器
encoder	[en'kəʊdə]	n.	编码器
calibration	[kælɪ'breɪʃ(ə)n]	n.	校准；刻度；标度
tachometer	[tæ'kɒmɪtə]	n.	转速计；流速计
contour	['kɒntʊə]	n.	轮廓
tolerance	['tɒlərəns]	n.	公差
Electro Discharge Machine (EDM)			电火花加工机床
Computer Aided Manufacturing (CAM)			计算机辅助制造
Initial Graphic Exchange Specification (IGES)			初始图形交换规范
postprocessing			后处理

Chapter 2
Basis of UG NX CAM

Objectives:
✓ To understand the terminology of CAM.
✓ To be able to create CAM operations and generate operation tool paths.
✓ To understand the postprocessor.

2.1 UG NX CAM overview

2.1.1 Brief introduction about UG NX CAM

NX provides complete computer-aided manufacturing (CAM) software solutions for machine tool programming, postprocessing and machining simulation. NX CAM software's advanced functions in each of its modules can maximize returns on your investments in the latest machine tool technology.

(1) Advanced programming capabilities.

NX CAM software provides a wide range of functionality, from simple NC programming to high-speed and multi-axis machining, enabling you to address many tasks with one system. The flexibility of NX CAM software means that you can easily complete the most demanding jobs. Fig.2-1 shows the advanced programming capabilities.

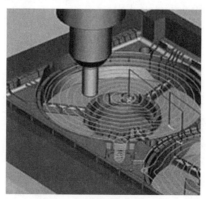

Fig.2-1 Advanced programming capabilities

18

(2) Programming automation.

The latest NC programming automation technologies (see Fig.2-2) in NX CAM software can improve your manufacturing productivity. With feature-based machining (FBM), you can reduce programming time by as much as 90 percent. In addition, templates enable you to apply pre-defined, rules-driven processes to standardize and speed your programming tasks.

Fig.2-2 Programming automation

(3) Postprocessing and simulation.

NX CAM software has a tightly integrated postprocessing system that enables you to easily generate the required NC code for almost any type of machine tool and controller configuration. Multiple levels of NC program validation include G-code-driven simulation, which eliminates the need for separate simulation packages. Postprocessing and simulation is indicated in Fig.2-3.

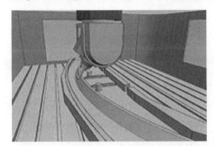

Flg.2-3 Postprocessing and simulation

(4) Integrated solution.

NX provides advanced CAD tools that can be used for everything from modeling new parts and preparing part models for CAM, to creating setup drawings directly from 3D model data. Fig.2-4 shows integrated CAD/CAM solution.

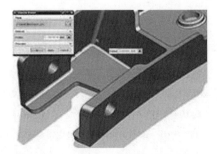

Fig.2-4 Integrated CAD/CAM solution

(5) Programming productivity.

You can easily access the advanced NC programming functions in NX CAM software. For example, the latest user interaction techniques and intuitive graphical driven programming enable you to quickly create NC programs. Fig.2-5 shows the programming productivity.

2.1.2 Terminology

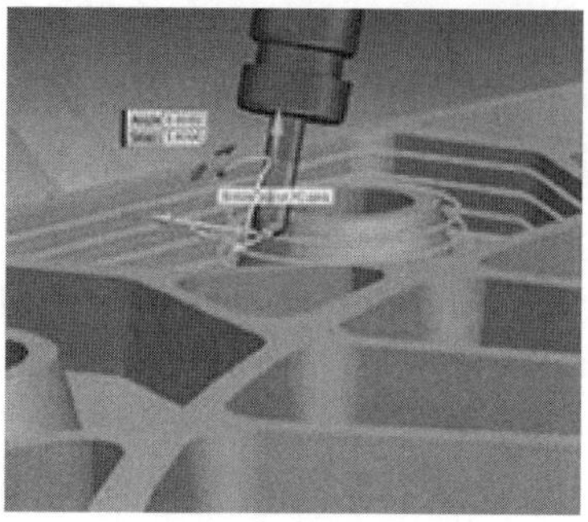

Fig.2-5 Programming productivity

An understanding of the following terms is essential to effective usage of the manufacturing application.

1. Setup

The setup is a part that can contain, or can be referred as components, the part to machine, blank, fixtures, clamps, and machine tool.

2. Assemblies

Assembly parts can be machined using NX manufacturing applications. You can select geometry in an assembly part file or any component part file to use in an operation. If the selected geometry is located in the component part file, the CAM operation will contain an occurrence of the selected geometry. All selected geometry is associative (See Associativity).

CAM objects (operations, tools) can only be retrieved in the assembly part file. You can retrieve libraries of tools or a component library part file. Part merge can be used to retrieve CAM objects from the component part into the assembly part.

An assembly can be created containing components, such as clamps and fixtures.

· This approach avoids having to merge clamp, fixture, etc. into the part to be machined.

· This approach allows you to generate fully associative tool paths for models that you may not have write access privilege.

· This approach enables multiple NC programmers to develop NC data in separate files simultaneously.

The manufacturing application retains the information used in generating a tool path. This capability is termed associativity.

3. Operation navigator

The operation navigator is a graphical organization aided with a tree structure that illustrates relationships between the geometry, machining method, and tool parameter groups and the operations within a program. Parameters are passed down or inherited from group to group and from group to operation based on their location in the tree structure. You can view and manage relationships between operations and parameter groups to share the parameters among operations.

4. Operation

An operation contains all information used to generate a single tool path.

5. Associativity

If the geometry or tool used for an operation is edited after generating its tool path, the operation automatically uses the new information when you regenerate. It is not necessary to re-select the geometry. If geometry required to generate the tool path is deleted, the software prompts you to specify new geometry.

If tool paths are automatically updated once, the geometry is altered, it can reduce productivity, since some operations take several minutes to complete. Therefore, it is up to you to update the tool paths when necessary. Similarly, when you specify a clearance line in turning, the coordinate data is saved, but modification of the line is not reflected in the operations.

6. MCS

There are two coordinate systems which are unique to manufacturing: the Machine Coordinate System (MCS) and the Reference Coordinate System (RCS).

The MCS is the base location for all output points of subsequent tool path. This allows you to move the Work Coordinate System (WCS) independently. If you move the MCS, you re-establish the base location for the output points of subsequent tool path.

7. Parent group

Parent group stores parameters that you use more than once outside of the operations, which helps you to reuse and organize the information required.

2.1.3 Manufacturing programming workflow

The Manufacturing module of UG NX allows you to interactively program tool paths, which includes the following paths.

· A default user interface that contains common functionality for milling, drilling, turning, and wire EDM operations.

· The ability to customize your user interface to meet industry specific requirements with templates which include machine tools, cutting tools, machining methods, shared geometry, and sequences of operations.

· An organization aid, the operation navigator, allows you to view and manage relationships between operations, geometry, machining methods, and tools.

Preliminary planning and analysis for manufacturing is shown as follows:

· Visually inspect the part to be machined or review the drawing.

· Determine the machining type or machining positions required.

· Analyze the part to collect data for the manufacturing plan.

· From the part analysis, determine the best tools and operation types to select tools, then create a rough plan and a finishing plan.

The manufacturing programming workflow is shown as Fig.2-6.

· Create a part to contain the setup, which has all the manufacturing information.

· This setup part can contain, or can be referred as components, the part to machine, the blank, fixtures, clamps, and the machine tool.

· Establish the program, tool, method, and geometry parent groups to define parameters for reuse.

· Create operations to define the tool paths.

· Generate and verify the tool paths.

· Post process the tool paths to format the data for your machine tools and controllers.

· Create shop documentation.

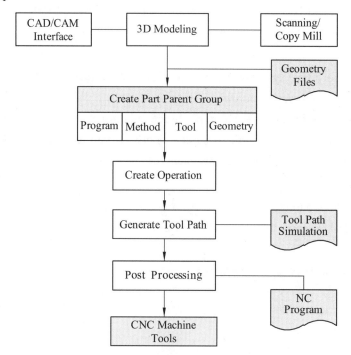

Fig.2-6 Manufacturing programming workflow

2.2 UG CAM machining environment

2.2.1 Manufacturing setup overview

The setup process creates a manufacturing assembly and adds data related to machining your part. With a setup you save the full machining environment.

· The setup is initialized from pre-defined setup parts (templates) in a CAM session. The template selected determines which operation types and subtypes are available.

· The setup stores all of your CAM data in a file. The setup can be in the same file as the part you are cutting, or in a separate assembly file.

· The setup consists of all operations, their resulting tool paths, and the environment in which the operations are created. This includes programs and parameters for tooling, geometry and cutting methods.

· The setup promotes standardization as it can be saved as a template and reused.

The setup starts with a part model. The requirements vary with part complexity. Simple parts may not need additional geometry for programming purposes. You can add a setup to the part file to create a simple setup file (non-assembly). More complex parts may need a fixture to add clarification.

2.2.2 Creating a setup in the file containing the part to machine

You can create a setup in the file containing the part to machine by the following steps.

· Choose Start→Manufacturing.

· In the Machining Environment dialog box, CAM Session Configuration group, select a configuration from the list.

· In the CAM Setup to Create group, select a setup from the list.

· Click OK.

Available CAM session configurations and setups are shown in Tab.2-1.

Tab.2-1   Available CAM session configurations and setups

CAM Session Configuration	Setups included
cam_express	all in the ascii library, general, mill, turn, mill_turn, hole_making, wedm, legacy, inch, metric, express, tool_building
cam_express_part_planner	all in the Teamcenter Manufacturing library
cam_general	mill_planar, mill_contour, mill_mulit-axis, drill, machining_knowledge, hole_making, turning, wire_edm, solid_tool. probing
cam_library	all in the ascii library, general, mill, turn, mill_turn, hole_making, wedm, legacy, inch, metric, express, tool_building
cam_teamcenter_library	all in the teamcenter manufacturing library
mill_contour	mill_contour, mill_planar, drill, hole_making, die_sequences and mold_sequences
mill_multi-axis	mill_multi_axis, mill_contour, mill_planar, drill and hole_making
mill_planar	mill_planar, drill and hole_making

To change the configuration for an existing setup, do as follows.

· Choose Preferences→Manufacturing.

· In the Manufacturing Preferences dialog box, click the Configuration tab.

· Click Browse (Configuration File).

· Select a Configuration File, for example, mill contour.dat, from the file list and click OK.

· Click OK to exit the Manufacturing Preferences dialog box.

2.3 Analyzing the part to machine

2.3.1 Manufacturing part analysis overview

Before you can create a manufacturing plan, you must analyze your part to collect the relevant data. The manufacturing plan would include key factors below.
(1) Machine type and machine positions.
(2) Cutting tools.
① Number of tools.
② Type of tools, for example, drill, reamer, mill.
③ Cutting tool parameters.
(3) Roughing and finishing plans.
① Operations.
② Operation sequence.
③ Stock requirements.

The data required to create the manufacturing plan would include information to specify the correct cutting tools, tool parameters and information to specify the correct stock requirements, such as the material type, and required tolerances.

2.3.2 Find part information for manufacturing

The part information for manufacturing can be found by several ways.
· Measure the distance.
· Analyze the part with NC assistant.
· View information for the selected geometry.
· Analyze the curvature.

Use the NC assistant to collect the data is necessary to machine your part, the process is illustrated in Fig.2-7. You can determine:
· Corner radius between walls using the corners option. Use this option to help select tool diameters.
·Corner radius between floors and walls using the blends option. Use this option to help select mill tool corner radius.
· Wall taper angles using the draft option. Use this option to help select mill tool taper angles.
· Cut depths using the levels option. Use this option to help select tool lengths, and to group common levels together for cutting.

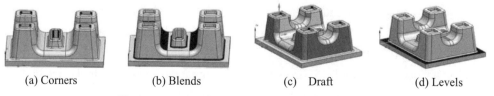

(a) Corners　　　　(b) Blends　　　　(c) Draft　　　　(d) Levels

Fig.2-7　Use the NC assistant to collect the data

The example shows how to determine tool lengths using the NC assistant command.

① Open a file that contains the part geometry (see Fig.2-8).
② Choose Start→Manufacturing.
③ Choose Analysis→NC Assistant.
④ Orient your geometry to a view that allows you to select the faces to be analyzed.
⑤ In the NC Assistant dialog box, Faces to Analyze group, click Select Faces.
⑥ In the Graphics window, select the Faces to be analyzed (see Fig.2-9). The entire blank stock and part are selected.
⑦ In the Analysis Type group, from the Analysis Type list, select Levels. Because this option recognizes the depths of all planar levels in the part, it helps you to identify the correct length of the tools for machining the part.

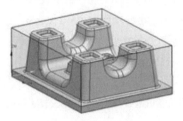

Fig.2-8   Find part information for manufacturing   Fig.2-9   Select the faces to be analyzed

⑧ In the Reference Vector group, from the Specify Vector list, select ZC Axis.
⑨ In the Reference Plane group, from the Specify Plane list, select Inferred.
⑩ In the Graphics window, select the Face to be referenced as a plane that is equal to zero (see Fig.2-10).
⑪ In the Limits group, enter values for minimum level and maximum level.
⑫ In the Tolerances group, enter values in the Distance and Angle boxes. These values set the tolerance range within which the software recognizes geometry.
⑬ In the Actions group, click Analyze Geometry to display colors on the faces that were analyzed (see Fig.2-11).

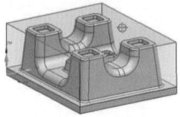
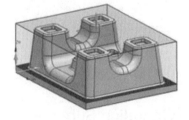

Fig.2-10   Select the face to be referenced   Fig.2-11   Display colors on the faces

⑭ In the Results group, select the Save Face Colors on Exit Check box. This ensures that the analysis colors on the faces are saved with the part.
⑮ In the Results group, click Information to view additional data on the faces that were analyzed.
⑯ Click OK.

2.4 Creating CAM operations

2.4.1 Operations overview

An operation generates one tool path, containing the information required to generate the tool path: geometry, tool, and machining parameters.

For each operation that you generate and accept, the software saves the information used to generate the tool path in the current part. This information includes the postprocessor command set, display data, and the definition of coordinate system.

You could specify all the required information within each operation. However, it is more efficient to specify parameters once and store the information where it can be used by multiple operations. Store only the most specific parameters, such as engage, retract, and stepover, in the operation. For example, To repeat similar operations on two cavities, you can put the cavity geometry in two groups, and then create or copy the same operations under each group. This saves you from duplicating your efforts.

The Create Operation dialog (as shown in Fig.2-12) enables you to create a specified operation type that uses parameters from four selected groups: type, operation subtype, location and name.

Fig.2-12 The Create Operation dialog

2.4.2 Operation Subtypes

All operations use a processor to calculate the tool path. NX provides operation templates for the various processors, including: Drill, Hole making, Turning, Wire EDM, Milling.

Milling is the most important and frequently-used processor. According to the shape of the machined surface, milling operation can be divided into planar milling and contour milling, as shown in Fig.2-13.

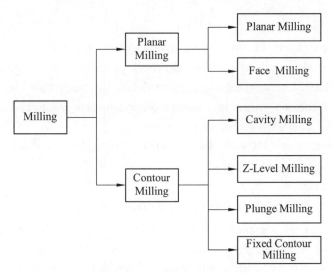

Fig.2-13 Milling operation subtype

Milling operation subtypes are show in Tab. 2-2.

Tab. 2-2 Milling processors and operation subtypes

Icon	Subtypes	Description
	Planar milling	follows 2D boundaries to remove material along vertical walls, or walls that are parallel to the tool axis. The areas on the part to be machined include planar islands and planar floors that are normal to the tool axis
	Face milling	cuts planar faces on solid bodies
	Cavity milling	removes large volumes of material. Cavity milling removes material in planar levels that are perpendicular to a fixed tool axis. Part geometry can be planar or contoured. Cavity milling is ideal for rough-cutting parts, such as dies, castings, and forgings
	Plunge milling	cuts vertically to take advantage of the increased rigidity when a tool moves along the Z-axis. Plunge milling efficiently roughs out large volumes of blank, and lets you finish hard-to-reach deep walls with long slender tool assemblies
	Z-Level milling	contours a part or cut area. Z-level milling removes material in planar levels that are perpendicular to a fixed tool axis. Part geometry can be planar or contoured
	Fixed-axis surface contouring	contours a part or cut area. Fixed axis surface contouring removes material along the part contours
	Variable-axis surface contouring	contours a part or cut area. Variable-axis surface contouring removes material along the part contours

2.4.3 Parent Groups

Storing parameters that you use more than once outside the operations, parent groups help you to reuse and organize the information required to machine your part. Parameters specified for a group are passed down from group to group and from group to operation. An operation retrieves the information required to generate the tool path from each of the following locations.

1. Geometry

Geometry groups define the machining geometry and the orientation of the part on the machine tool. Parameters such as part, blank, and check geometry, MCS orientation, and clearance planes are defined here. Use geometry groups to define geometry that can be shared between multiple operations. Geometry groups may contain other geometry groups and cutting operations.

The geometry for a part defines following conceptions.
· The orientation, which includes the coordinate system, fixture offset, clearance plane and tool axis. (These are stored in a MCS group.)
· The areas to machine.
· The part material, which is used to calculate machining data.

Geometry selections in a manufacturing setup represent either the part model that requires machining, portions of the part model, or additional geometry added by the programmer.
· Blank geometry represents the raw material and can be cut through or engaged into directly.
· In Process Workpiece (IPW) geometry represents the machined workpiece at the end of each stage of machining.
· Part geometry represents the finished part.
· Check geometry represents areas you do not wish to violate such as clamps that hold the part.
· Trim geometry constrains the cut regions.
· Cut area or face geometry define specific areas to machine.
· Wall geometry.
· Floor geometry.
· Geometry added by the programmer helps to control the tool path during the manufacturing process.

Depending on the operation type and subtype, the geometry you use may be a solid body, face, or boundary, as shown in Tab.2-3.

Tab.2-3 Milling geometry

Icon	Subtype	Description
MCS	MCS	Use the Milling MCS to specify: ①MCS ②RCS ③Clearance plane ④Lower limit plane ⑤Avoidance
	Workpiece	Use the Milling workpiece to specify: ①Part ②Blank ③Check ④Part Offset ⑤ Part material
	Mill area	Use Mill area to create specify: ①Part ②Check ③Cut Area ④Walls ⑤Trim boundaries
	Mill bnd	Use Mill bnd to create specify: ①Part boundaries ②Blank boundaries ③Check boundaries trim ④Trim boundaries ⑤Floor
A	Mill text	Use Mill text to specify: ①Drafting ②Text Floor
	Mill geom	Use the Mill geom to specify: ①Part ②Blank check ③Part offset ④Part material
	Drill geom	Use Drill geom to specify: ①Holes ②Part surface ③Bottom surface ④Tool axis

The Mill geometry and Workpiece icons in the Create Geometry dialog box (see Fig.2-14) perform identical functions. They both allow you to define part, blank, and check geometry from selected bodies, faces, curves, or surface regions. In addition, they allow you to define part offset, part material, and save the currently displayed layout and layer.

Fig.2-14　Create geometry dialog box

Trim boundaries apply to the area milling and flow cut drive methods of fixed axis surface contouring. Trim boundaries further constrain the cut regions.

Part offset adds to or is subtracts from the modeled part geometry by a specified offset (thickness) value. A positive value offsets the thickness outside the part. A negative value offsets the thickness inside the part. The cutter recognizes the part offset as the workpiece and then machines to it accordingly. Part stock and custom stock are measured from the part offset. This is useful when the model geometry is offset from the part you are machining, such as an EDM electrode or a sheet metal die.

Geometry parents are used to define geometry for manufacturing. You must first create the geometry parent, and then add geometry to the new geometry parent. The geometry can then be inherited by manufacturing operations or other geometry parents.

2. Tool

Tool groups define the cutting tools. You may create tool groups either by creating a tool from a template or by retrieving tools from a library.

You can specify mill, drill, and lathe tools, and save data associated with the tool to use as default values for the corresponding postprocessor commands. For example:

· Adjust and Cutcom register numbers.
· Tool/Face number.
· Spindle directions.
· Offsets.
· User-defined attributes.

Each operation needs an appropriate tool to machine the part area. Tool parameters define:

· The cutting insert shape and dimensions.
· The tool holder.

· The tool material, which is used to calculate machining data.

Tool parameters can be retrieved from a library with hundreds of standard tools, or created as needed. Each cutting tool is placed in a holder, within a carrier and turret, and loaded on a machine tool.

The tool Types (templates) available in the Create Tool dialog box are determined by the specified CAM setup (template part file). The Create button allows you to create a new tool based on the chosen tool subtype icon (template).

The Create tool dialog allows you to create tools used in various operation types, as shown in Fig.2-15.

Fig.2-15 Create tool dialog box

The Create tool dialog options are explained below.

· The types of tools that are available are determined from the machining environment parameters that were previously selected.

· The type determines the subtype or class of tools that are available for selection.

· Select the subtype and click OK or Apply.

· Enter parameters and click OK.

The following (see Tab.2-4) is a list of milling tools and the key dimensions that you can set for each tool.

Tab.2-4 Milling Tools

Icon	Subtype	Drafts	Parameter Description
	5 Parameter mill tool		(D) Diameter (R1) Lower radius (L) Length (B) Taper angle (A) Tip angle (FL) Flute length
	Ball mill		(D) Diameter (B) Taper angle (FL) Flute length (L) Length

Tab.2-4

Icon	Subtype	Drafts	Parameter Description
	Face mill		(D) Diameter (R1) Lower radius (L) Length (A) Tip angle (B) Taper angle (FL) Flute length
	7 Parameter mill tool		(D) Diameter (R1) Lower radius (L) Length (B) Taper angle (A) Tip angle (FL) Flute length (X1) X center R1 (Y1) Y center R1
	10 Parameter mill tool		(D) Diameter (R1) Lower radius (L) Length (B) Taper angle (A) Tip angle (X1) X center R1 (Y1) Y center R1 (R2) Upper radius (X2) X center R2 (Y2) Y center R2 (FL) Flute length
	Barrel cutter		(D) Diameter (R) Barrel radius (R1) Lower radius (R2) Up radius (L) Length (FL) Flute length (Y) Y center (SD) Shank diameter
	T cutter		(D) Diameter (R1) Lower radius (R2) Up radius (L) Length (FL) Flute length (SD) Shank diameter

When specifying a tool for an operation, you can create a tool group by retrieving a tool from the cutting tool library (defined by the configuration file), which contains much more information about tools than is required to generate tool paths. The description from the library is mapped to the template subtype, and can be displayed in the operation navigator. If the other library information is required for shop documentation or postprocessing, you can retrieve it directly from the library at that time.

The system outputs the tool number located in the operation and displayed in the operation navigator to the postprocessor or CLSF file. The tool number (or T code) can come from three places: the pocket, the tool, and the operation. The tool number is inherited from the pocket level to the tool level to the operation level. Based on what the T code means on your NC machine, you need to decide which is the best way for you to assign tool numbers in your programs.

3. Method

Method groups define the type of cut method (rough, finish, semi-finish). Parameters such as Intol, Outol, and Part Stock are defined. The various parameters used to define mill, drill, lathe, and wire EDM method groups are described in the following sections: feed rates, specify colors, additional passes, part stock, Intol/Outol, display options, inheritance.

A machining method defines other parameters to share with multiple operations.

· Stock, tolerances, feed rates and display colors.

· The cut method, which is used to calculate machining data.

The Create Method dialog allows you to create machining method parent groups used in various operation types. The types available are determined from the machining environment parameters that have been selected. The type determines the subtype of methods that are available for selection.

The Create Method dialog (see Fig.2-16) options are explained below.

Fig.2-16　Create Method dialog box

· The Type drop-down list allows you to select the type of machining for this method parent.

· The Type determines the subtype, a more specific type of machining, for in this method parent.

· The parent group drop-down list box allows you to select previously defined method groups. The method you are creating is assigned to the method group you select.

· The name field allows you to choose your own name for the method group, helping you to organize your program in a way that works for you.

After setting the create method dialog options (type, subtype, parent group, and name), selecting OK displays the parent group type dialog (as shown in Fig.2-17), based on the subtype selection. For example, if your subtype selection was mill method, then the mill method dialog would be displayed. You would then select the various options that would be assigned to the parent group, mill method.

Fig.2-17　Parent group type dialog

· Part stock is the amount of material to be left on the part after machining. This value will be inherited by operations that are created under the method parent. You can override the inherited value by entering a new value in the operation.

· Intol and Outol define an allowable range that the tool may use to deviate from the part surfaces. The smaller the values, the more accurate the cut. Intol allows you to specify the maximum amount by which the tool may penetrate through the surface. Outol allows you to specify the maximum amount by which the tool may avoid contacting the surface.

· Feed rate values can differ form several types of tool movement during the progress of the tool path.

· The specify colors (see Tab.2-5) option allows you to display each different type of tool motion in an individual color to help in visual validation of the tool path avoidance control and feed rates.

Tab.2-5　Specify colors

Motion Types	Colors
Rapid	Red
Approach	Blue
Engage	Yellow
First Cut	Cyan
Stepover	Green
Cut	Cyan
Traversal	Blue
Retract	White
Departure	Blue

Many of the CAM setups include predefined methods. You can use them in default settings, or you can customize. For example, the mill contour template has methods of Mill Rough, Mill Semi-Finish, and Mill Finish. You can create new operations in these groups to inherit different stock values.

Some parameters are passed down from group to group and from group to operation. For example, an operation or a method group inherits display and tolerance parameters from another method group. Not all operations can be placed in all method groups. Different operations can inherit different parameters. Method inheritance can be overridden in an operation.

4. Program

Program groups enable you to group and order operations into programs. For example, all operations required to machine the top view of a part may compose a program group. By grouping operations, you may output many operations at once in their proper sequence by selecting the program group and then choosing tools>operation navigator>output>CLSF or postprocess.

The Program Order view of the Operation Navigator shows which program group each operation belongs to and the order in which operations will be executed on the machine tool. This is the only view of the operation navigator in which the order of the listed operations is relevant or important.

· A program is equivalent to a tape file.

· A program usually contains multiple operations and generates all tool paths for these operations.

· A program specifies the order to run the operations.

· A program requires postprocessing to create the actual commands that are sent to the CNC controller.

You can create multiple programs for each job, such as roughing program, finishing program and drilling program. Before postprocessing, you can reorder the operations in each program, or move them from one program to another.

The Create Program dialog (as shown in Fig.2-18))allows you to create program groups. These program groups can be seen as containers for various operations. Program groups become program parents when you place operations in them in the operation navigator.

Fig.2-18 Create program dialog

The Create Program dialog options are explained below.

· The type drop down list allows you to select the template part that you are using. Most of the template parts contain program templates.

· The subtype is the type of program group that "programs" are used to group operations for output.

· The parent group drop down list box allows you to select the previously defined parent groups. The system creates the program in the parent group you select.

· The name field allows you to select your own name for the program group, helping you to organize your program in a way that works for you.

· The program parents, which you have created in the Create Program dialog, are used to specify the order of operations for output of CLSF files and/or postprocessing. It is a suggested practice to have numerous program parent Groups created, naming them according to the programs that are being output. When you are ready to generate your output, selection of the program parent selects all of the operations assigned to that parent group.

2.5 Coordinate system of UG NX CAM

There are different coordinate systems in NX CAM. The most frequently used coordinate systems for manufacturing are the following.

(1) Absolute Coordinate System (ACS).
(2) Work Coordinate System (WCS).
(3) Machine Coordinate System (MCS).
(4) Reference Coordinate System (RCS).

2.5.1 Absolute Coordinate System

The ACS is not visible or movable. The direction of the global coordinate system axes are the same as the view triad, but not its origin. ACS is convenient to look for the relationship between the components in assembling.

2.5.2 Work Coordinate System

The ACS is movable in space, which is used for modeling and manufacturing widely. It determines most input parameters, such as from points, clearance planes, and the tool axis. In case of requirement of transforming coordinate system, such as rotation and translation, WCS can be introduced to control the modeling coordinate values conveniently, so as to reduce the system overhead of matrix transformation.

2.5.3 Machine Coordinate System

The MCS determines the orientation and origin of tool paths for all operations in the orient group. The MCS (see Fig.2-19) has an initial position that is matched to the ACS. The position of the MCS is saved in the part file. Each orient group (for example, MCS, MCS MILL) defines the MCS that is necessary to machine a specific side of the part. If you move the MCS, you reestablish the base location for the subsequent tool path output points. The MCS has the following characteristics.

· It stores the RCS.

· It can store the clearance plane, lower limit plane, and avoidance points.

· Operations stored under the MCS parent inherit the parameters specified in the MCS parent.

· Operations moved from one MCS to another do not need to be generated to reflect the new orientation or origin.

The MCS is useful when you do the following.

· Output the tool path in relation to the machine home position or any other frequently used setup location.

· Reorient the machine tool axes to the workpiece.

· Shift large parts for successive tool paths.

· Set up parts which no longer have a reference point. For example, a tooling hole which has been machined away.

· Reestablish setup position and orientation after rotary table movement and after compound axis movement.

· Maintain critical dimensions otherwise lost to warping or tolerance buildup.

· Establish basic and true positions.

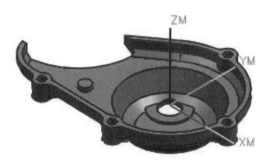

Fig.2-19 Machine Coordinate System

The MCS references the NC/CNC program origin which is the zero reference point for all tool path output. If the MCS moves, the zero point of all the tool paths that use it move with it. It is helpful to create and save a clearance plane when you specify the orientation and origin of the machine coordinate system in an MCS parent group (as shown in Fig.2-20). The clearance plane defines a safe clearance distance for tool motion before and after an operation.

Fig.2-20 Mill orient (MCS) dialog box

2.5.4 Reference Coordinate System

RCS is the coordinate system that maps parameters when you create, copy, move or transform an operation. By default, it is in the same location as the ACS, whose coordinates are $X = 0$, $Y = 0$, and $Z = 0$. You can specify the RCS either when you create or edit the MCS. Every orient group has an RCS.

You can use the RCS to relocate tool axis vectors, clearance planes, and avoidance points after transforming an operation.

Specify an RCS to help you relocate tool axis vectors, clearance planes, and avoidance points after transforming an operation.

(1) In the operation navigator, in the geometry view, double click the MCS.

(2) In the Mill orient dialog box, in the RCS group, specify the RCS location in one of these ways:

① To place the RCS in the same location as the MCS, select the link RCS to MCS check box.

② To place the RCS in a different location from the MCS, clear the link RCS to MCS check box, and then specify the RCS.

You can set a preference to always link the RCS to the MCS when you create a new MCS.

(1) In the Manufacturing application, choose preferences→manufacturing.

(2) In the Manufacturing preferences dialog box, click the operation tab.

(3) In the operation property page, under edit, select the link MCS/RCS check box. Click OK.

2.6 Tool Paths

2.6.1 Generating Operation Tool Paths

There are several ways in which you can generate an operation tool path. You can generate tool paths for:

(1) A single operation from the operation dialog box.

(2) One or more operations from the operation navigator, operations toolbar, or tools menu. You must wait for the operations to finish generating before you can continue working.

(3) One or more operations in the background as a parallel generate process from the operation navigator, operations toolbar, or tools menu while you continue working in the current part file.

(4) A single operation or group in the background as a batch process. You must close the part file associated with the batch process. You can work in another part file while the batch process runs. You can reopen the part file associated with the batch process only after the batch process is complete.

Use the parallel generate command to generate operations in the background while you continue working. Select any number of operations at the same time, and they count as one parallel process. You can:

(1) Run multiple processes concurrently.

(2) Stop parallel processes with the stop parallel generate command.

Batch process generates tool paths, postprocess groups and/or operations, and generates shop floor documentation. You can process all three outputs at one time, or create each output individually. You can use batch processing to:

(1) Generate multiple outputs from one command.

(2) Work on another project while the process takes place.

(3) Schedule the process for the most convenient time, or when the system is most free. You do not need to be present.

While parallel generating, the part must remain open in the current session. You can continue to work on it, or work on another part, but you cannot close the part. This is different from batch process, which requires you to close the part and to wait until the batch processing is complete to open the part.

2.6.2 Displaying Tool Paths

Use the edit display command to specify how the tool, tool holder, and tool path are displayed in the graphics window (as shown in Fig.2-21).

The availability of the process display parameters options varies depending on the machining module you use.

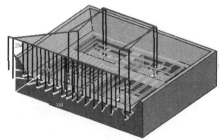

Fig.2-21 Displaying tool paths

Display options are shown in Tab.2-6.

Tab.2-6 Display options

Option		Description
Tool	Tool display	①None ②2D ③3D ④Axis
	Frequency	Sets the frequency of the tool display
	Pattern	Enables tool holder to display during tool path generation and replay. You can design a tool holder in the modeling application, save it as a pattern, and then use it in an operation to visually check for interference with the part during machining
Path	Path display colors	Lets you set the color of the display of each type of tool motion to help in visual validation of the tool path
	Speed	Sets the rate of the tool display movement. You can move the slider to set a relative value between 1 and 10. A speed of 1 is the slowest; a speed of 10 is the fastest
	Path display	Sets how the tool path is displayed in the graphics window: ①Solid ②Dashed ③Silhouette ④Fill ⑤Silhouette Fill
	% tool	Sets the percentage of the tool that is engaged in cutting
	Path normal	Sets the output of a silhouette normal to the tool axis or to a user-specified vector. Choose between the two options, Tool Axis and I,J,K
Path Generation	Display Cut Region	Available in Planar and Cavity Milling only. Displays the curve defining the machinable region or regions for each cut level before processing the tool path
	Pause After Display	Available in Planar Milling, Cavity Milling, and Fixed and Variable Contour operations only. Pauses the processor after displaying the machinable region or the tool path at each cut level
	Refresh Before Display	Available in Planar Milling, Cavity Milling, and Fixed and Variable Contour operations only. Removes all existing temporary screen displays before displaying the generated tool path
	Suppress Path Display	Does not display the generated tool path

Pattern allows you to display a tool holder during tool path generation and replay. You can design a tool holder in the modeling application, save it as a pattern, and then use it in the operation to visually check for interference with the part during machining.

2.6.3 Verify Tool Path

Use Verify command to verify your tool path by playing an animation of the tool path or material removal. Animating material removal helps you verify that the cutter is cutting the specified portions of the raw material during tool path generation.

The Tool path visualization dialog box opens when you choose this command, as shown in Fig.2-22.

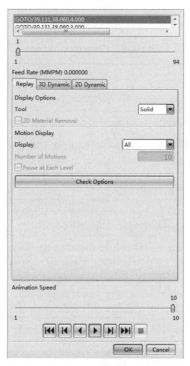

Fig.2-22　Tool path visualization dialog

Tab. 2-7 lists options in the main dialog box, and options that are common to the three tabs: Replay, 2D Dynamic, and 3D Dynamic.

Tab.2-7　Tool path visualization dialog options

Options		Icon	Description
Graphical display window		—	Displays the tool path or material removal in the graphical display window. Replay allows selection of tool path points in the display
Path listing		—	Lists the tool path for the operation you are replaying. When using the replay option, if you select a point in the graphical display window that point will be highlighted in the listing window of the tool path visualization dialog
Feed rate		—	Displays the feed rate for the current move
Animation speed		—	Controls the speed for the tool path replay
Animation control	Rewind to previous operation	⏮	Resets the replay to the start of the path or the previous operation
	Step backward	⏴|	Moves the animation back by one tool path position
	Reverse play	◀	Replays the tool path in reverse order
	Play	▶	Plays the current tool path
	Step forward	|▶	Moves the animation forward by one position in the tool path
	Next	⏭	Moves the animation to the next operation. In multi-level tool paths, the animation is moved to the next level
	Stop	■	Stops the tool path motion (Only in 3D Dynamic)

1. Replay

Use the replay options in the Tool path visualization dialog box to view a replay of the NC program. You can view the cutter at each program location. Because a replay does not include material removal, it is the fastest of the three animation techniques that are available in the tool path visualization dialog box.

You can use the replay options to:
- Display the tool or tool assembly for one or more tool paths.
- Display the tool as a wire frame, a solid or a tool assembly.
- Display gouges if they exist and also view a report on the gouges.
- Control the tool path display.

2. 3D Dynamic

3D Dynamic material removal displays the moving tool and tool holder along one or more tool paths showing the material being removed. This mode also allows you to zoom, rotate, and pan in the graphic window. Blank geometry is used to represent the rough material or raw stock.

You can use the 3D Dynamic options to:
- Display the tool or tool assembly for one or more tool paths.
- Display the tool as a wire frame, a solid or a tool assembly.
- Display gouges if they exist and also view a report on the gouges.
- Check for collisions.
- Control the tool path display.

Before you can create an IPW color plot, you must have a 2D Dynamic or 3D Dynamic model. You can generate these models when you play an animation using either the 2D Dynamic or 3D Dynamic options in the Tool path visualization dialog box.

3. 2D Dynamic

2D Dynamic material removal displays the tool path display including material removal. Use the 2D Dynamic options in the Tool path visualization dialog box to view material removal as well as the tool path.

You can use the 2D Dynamic option to:
- View animated dynamic material removal for single or multiple operations.
- Create faceted bodies for IPW, gouges and excess material.
- Use the faceted bodies for input in subsequent operations.
- Verify milling and drilling operations.
- View the display of the blank and the dynamic removal of material in a pixel-based view.

2.7 Postprocessing

2.7.1 Postprocessing overview

The manufacturing application generates NX tool paths that are used to manufacture parts.

Generally, you cannot just send an unmodified tool path file to a machine and start cutting because there are many different types of machines. Each type of machine has unique hardware capabilities and is controlled by a specific computer (also called the controller).

The controller accepts a tool path file and directs tool motion and other machine activity (for example, turning the coolant or air on and off). Just as each type of machine has unique hardware characteristics, each controller has unique software characteristics. For instance, most controllers require a particular code to turn on the coolant. Some controllers also restrict the number of M codes that are allowed in one line of output. This information is not in the initial NX tool path.

Without the correct formatting for a machine, the tool path file hits the controller's brick wall of incompatibility as thown in Fig.2-23.

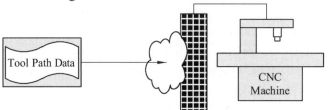

Fig.2-23 Tool path file hit the controller's brick wall

Therefore, the tool path must be modified to suit the unique parameters of different machine/controller combination. The modification is called postprocessing, and the result is a postprocessed tool path.

Two elements are essential for postprocessing. They are:

(1) Tool Path Data: This is an NX tool path.

(2) A Postprocessor: This is a program that reads the tool path data and reformats it for use with a particular machine and its accompanying controller.

The tool path data is reformatted by the postprocessor for the machine. Each postprocessor program is usually dedicated to a single type of machine/controller combination. You can modify postprocessor file parameters for functions of that particular type of machine/controller combination. However, you cannot modify the program for use with another type of machine/controller combination.

NX provides a generalized postprocessor program, Post, which inputs NX tool path data, and outputs machine readable NC code. Post is highly customizable and can be used for both very simple and very complex machine tool/controller combinations.

Post Builder is the NX product that is used to customize the postprocessor for each machine tool/controller combination.

2.7.2 The Postprocessor

NX provides the postprocessor that can properly format tool paths for specific types of machine/controller combinations. The postprocessor requires several elements, as shown in Tab.2-8.

Tab.2-8 The Postprocessor elements

Element	Description
Event generator core module	Cycles through the events in a part file and communicates the data associated with each event to the postprocessor. An event is a collection of data, that when processed by Post, causes the NC machine to perform some specific action. This is activated by following the path Tools→Operation Navigator→Output→Postprocessor, or the icon
Event handler (.tcl) file	Contains a set of instructions dictating how each event type is processed. This is created with Post Builder
Definition file (.def) file	Contains static information related to a particular machine tool/controller combination. This is created with Post Builder
Output file	Contains the postprocessed NC instructions that will be read and executed by the machine tool
Post user interface file (.pui)	Used by post builder to edit the event handler and definition files

The event generator, the event handler, and the definition file are dependent upon each other. Together they transform the tool path data contained in the part file into a set of formatted instructions that can be read and executed by a specific machine tool/controller combination.

The Postprocessor (see Fig.2-24) does the following:

(1) Use the event generator to read the events (tool path data) in the part file.

(2) Each event is processed according to the instructions contained in the event handler.

(3) The resulting instructions are formatted according to the information contained in the definition file.

(4) The postprocessed machine control instructions are written to the output file.

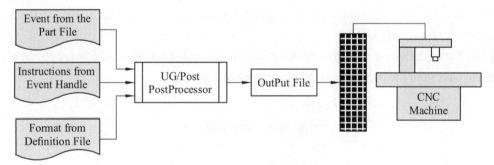

Fig.2-24 The Postprocessor

The tool path data is postprocessed according to the instructions in the event handler and the formats in the definition file.

To develop a post, you must create an event handler and a definition file. The recommended method to create these is with post builder. After creating a post, you will have three files: <post_name>.tcl, ~.def, and ~.pui.

Tab.2-9 is a brief description of the sample postprocessors. For production you should customize these postprocessors or create your own postprocessors with post builder. Any postprocessor should be thoroughly tested before running on the actual machine. All these postprocessors output inch units.

Tab.2-9 The Sample Postprocessor

Postprocessor	Description
wire_edm_2_axis	Two axis wire EDM machine with Mitsubishi control
mill_3_axis	Three axis vertical mill with Fanuc control
mill_4_axis	Four axis horizontal mill with a B axis rotary table and Fanuc control
mill_5_axis	Five axis mill with A and B axes rotary tables and Fanuc control
lathe_2_axis_tool_tip	Lathe programmed with the tool tip
lathe_2_axis_turret_ref	Lathe programmed with the turret reference point
millturn	Mill/Turn machining center with XYZ or XZC motion, lathe tool tip, and Fanuc control, changes the mode based on the operation type
millturn_multi_spindle	Machining center with Fanuc control links three postprocessors, lathe_2_axis_tool_tip, an XZC Mill with Z axis spindle, and an XZC mill with X axis spindle. It changes postprocessors based on the head UDE
tool_list (text)	Generates a list of tools used by the selected program in text format
tool_list (html)	Generates a list of tools used by the selected program in html format
operation_list (text)	Generates a list of operations in the selected program in text format

2.7.3 Postprocess an operation or program

The Postprocessor dialog box is shown in Fig.2-25. From the available machines list, you can select the machine that you want to create a post for. If the required machine is not listed, add it to your manufacturing configuration.

Units list specifies the units to output for the postprocessed coordinates. Some options convert coordinate data from the part units to the postprocessor units.

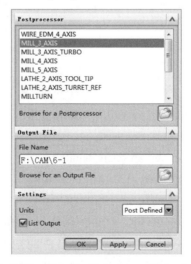

Fig.2-25 The Postprocessor dialog box

The following steps show how to postprocess an operation or program.

(1) In the operation navigator, choose the program or operation you wish to postprocess or choose the root in the program order view to postprocess all operations.

(2) Choose postprocess in the main menu bar or Tools→Operation Navigator → Output → NX postprocess.

(3) In the Postprocess dialog, select an available post from the list of sample postprocessors. You can use the Browse button beneath the list of posts to select postprocessors in any location. Browse can only be used with post builder created posts (.pui).

(4) Then specify an output file destination and name. This is where you store your output. You can turn on the list output option to see the output in the information window, in addition to the output to a file.

(5) Click OK. The output file is shown in Fig.2-26.

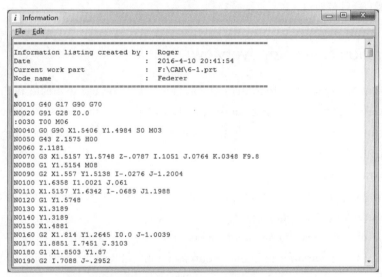

Fig.2-26 The postprocessor output file

2.7.4 Output CLSF

After an operation is generated and saved, the resulting tool path is stored as part of the operation within the part file. Output CLSF exports internal tool paths to a Cutter Location Source File (CLSF) for use by Graphics Postprocessor Module (GPM) and other postprocessors. The GOTO values exported are referenced from the MCS stored in the operation. With customization you can modify these formats by changing their Event Handler (.tcl) and Definition (.def) files for specific requirements.

To output a CLSF for an operation or program, follow the list below:

(1) In the Operation Navigator, choose the program or operation you wish to output or choose the root in the Program Order View to output all operations.

(2) Choose the Output CLSF icon in the main menu bar or Tools→Operation Navigator→Output→ CSLF. This displays the Output CLSF dialog.

(3) In the dialog, choose an available format from the list.

(4) Then specify an output file destination and name. This is where you store your output.

(5) You can turn on the list output option if you would like to see the output in the information window, in addition to the output to a file. If you have only made minor changes to the program and do not want to take the time for the display to appear on the screen, you can uncheck list output to speed up the output process.

This produces the output file, which you provide as input to GPM or your third party postprocessor.

2.8　Operation Navigator

2.8.1　Operation Navigator overview

The Operation Navigator has four hierarchical views that you use to create and manage an NC program. Each view organizes the same set of operations based on the theme of the view (as shown in Fig.2-27): the order of operations within the program, the tools used, the geometry machined, or the machining methods used.

Use the Operation Navigator to:

(1) Cut or copy and paste operations, programs, methods, or geometry within the setup for a part or between setups for different parts.

(2) Drag and drop groups and operations within the setup for a part.

(3) Specify common parameters in one group location, such as the workpiece geometry group. The parameters are passed down (inherited) by operations within the group.

(4) Turn on inheritance for specific parameters.

(5) Display the tool path and geometry of an operation in the graphics window to quickly see what is defined and which areas are machined.

(6) Display the In Process Workpiece (IPW) of a milling or turning operation.

Name	Path	Tool	Geometry	Method
GEOMETRY				
Unused Items				
MCS_MILL				
WORKPIECE				
PLANAR_MILL	✓	MILL_D8	WORKPIECE	MILL_ROUGH
FINISH_WALL	✓	MILL_D8	WORKPIECE	MILL_FINISH

Fig.2-27　Operation Navigator view

Within the Operation Navigator, a tree structure controls the relationships between groups and operations. The position of a group or operation determines how parameters are passed down (inherited). Individual columns display different types of information. You can customize the general appearance and select the columns to display for each view with the Properties command.

In the Operation Navigator, you can:
- Copy a group or operation
- Move an operation to the top of a group.
- Copy an operation from one setup to another.
- Lock or unlock a tool path.
- Approve a tool path to override a Regenerate or Repost status.
- Remove the approval on a tool path.
- Specify time estimates for tool changes and rapid feeds.

2.8.2 Operation Navigator Columns and Status Indicators

The Operation Navigator displays icons along with, or instead of, text in some of the columns. For example, the Name column uses icons as well as text to communicate the name and status of the operation. When you point to an icon, a tool tip displays with a description of the icon and related information. For example, the Toolchange column tool tip includes the name of the new tool.

The operation status is displayed as Tab.2-10.

Tab.2-10 The operation status

Icon	status	Description
✔	Complete	The tool path has been generated and the output (postprocessed or CLSF) is up-to-date
⊘	Regenerate	The tool path for the operation has never been generated, or the generated tool path is out-of-date
❗	Repost	The tool path has never been output, or the tool path has changed since it was last output and the last output is out-of-date

The tool path status is displayed as Tab.2-11.

Tab.2-11 The Tool path Status

Icon	status	Description
✔	Generated	The tool path has been generated. It may or may not contain tool motions
✘	Regenerate	The tool path has either not yet been generated, or been deleted

2.8.3 Operation Navigator view

The Operation Navigator has four views which display the same set of operations differently, as shown in Tab.2-12.

Tab. 2-12 Operation Navigator view

Icon	View	Description
	Program order view	Organizes operations by the order in which they are executed on the machine tool. Each program group represents a separate program file that is output to the postprocessor or CLSF
	Machine tool view	Organizes operations by the cutting tools used and displays all the tools retrieved from the tool library or created in the current setup. You can also organize cutting tools by turrets on a lathe or by tool type on a mill
	Geometry view	Organizes operations by machining geometry and MCS orientation. Each geometry group displays operations by the order in which they are executed on the machine tool
	Machining method view	Organizes operations by common machining applications which share the same parameter values, such as roughing, semi-finishing, and finishing

Each view organizes the operations into groups that control relevant parameters. The contents of a group are inherited by the groups and operations below it in the Navigator tree. The groups display whether they are actually used in the NC program or not.

For example, if a tool is used, it has one or more operations below it in the machine tool view. Unused tools do not have an operation below the tool.

When you select an operation in a view, then change views, the groups are expanded to show the selected operation in each view. You can also look at the dependencies panel to see where an operation is located in all four views.

You can switch from one view of the Operation Navigator to another by clicking the appropriate button on the Navigator toolbar.

(1) Using the program order view to create your NC program.

When you have viewed the program order view and have seen the order in which the operations are going to be machined, you can analyze the situation and decide if it is arranged in the most efficient way. If you decide that some operations need to be rearranged to make the program more efficient, it is easy to rearrange the operations in the program order view. For example, if you are trying to reduce the frequency that tool changes you can either choose to change the order of the operations, or change the tool that the operation uses. Changing the order of the operations is most easily done in the program order view. Changing the tool the operation uses is most easily done in the tool view. This is because the program order view groups the operations together in chronological order and the tool view groups the operations by the tool.

(2) Using geometry view to enhance your NC programs.

If you want to plan your NC program based on geometry, use the geometry view instead of the program order view. You can then organize the setup by geometric feature.

For example, you can create a geometry group for pocket A that includes a rough mill operation and a finish mill operation. You do the same for pocket B. In the program order view, you change the order of the operation so that you rough both pockets first then finish both pockets. In the geometry view, your operations are organized by the geometry they machine, but in the program order view, those same operations are organized in the order they are output.

If you create your NC program in the program order view, you can still use the geometry view to help you edit the NC program. Use the geometry view to evaluate the operations used in each geometry group and decide if the arrangement is the most efficient. Since the geometry view groups the operations together according to their geometry group, the Geometry View can be used to easily move or copy an operation from one geometry group to another.

Always check how the changes you have made in each view affect the Program Order View, as this is the view that contains the sequence of the operations for output.

(3) Using the Tool View to Enhance Your NC Program.

When you have viewed the Tool View and have seen all the operations in which the tools are used, you can analyze the situation and decide if it is arranged in the most efficient way. Use this view to change the tool the operation uses, by moving the operation to a different tool parent. Since the tool view groups the operations together by tool name, it allows you to quickly locate the tool you are searching for. Therefore, making changes in your operation relating to the tool are most easily in the tool view. For example if you modified a tool, you would use the Tool View to see all of the operations that used that tool, and regenerate them to update their tool paths.

Fig.2-28 is an example of the Tool View. In the furthest left hand column, under name, you can see all the tools that are in the part. You can also see that two tools are actually used by operations.

Name	Path	Tool	Description	Tool Nu...	Geometry	Method	Order Group
GENERIC_MACHINE			Generic Machine				
Unused Items			mill_contour				
BALL_MILL			Milling Tool-Ball Mill				
FIXED_CONTOUR	✓	BALL_MILL	FIXED_CONTOUR		WORKPIECE	MILL_ROUGH	NC_PROGRAM
FLOWCUT_MULTIPLE	✓	BALL_MILL	FLOWCUT_MULTIPLE		WORKPIECE	MILL_ROUGH	PROGRAM
BALL_MILL_D2			Milling Tool-Ball Mill				

Fig.2-28 Tool view

(4) Using the method view to enhance your NC programs.

You can use the method view (see Fig.2-29) to analyze the situation and decide if it is arranged in the most efficient way. In the method view the system displays the operations grouped together according to their machining method (rough, finish, semi-finish). This organization makes it easy to manipulate the method choice within the operations. Always be sure to check how the changes you have made in any view affect the Program Order View as this is the view that contains the sequence of operations.

Use this view to change the method information for several operations at once. For example, to change the cut color for all Mill-Rough operations, edit the Mill-Rough Method, and change the display options. Now all the operations in that method inherit the new colors. This is easier than changing the display options in each operation.

In most cases, changing an operation changes its status to regenerate and you must regenerate to make sure the tool path is up-to-date. Examples of changes that require regenerating are tool, geometry, or cutting parameter changes.

Other changes only require reposting or generating the CLSF to update the output files. Examples of changes that do not require regenerating the tool path are: feed rates, post commands, or the MCS parent group.

Fig.2-29 Method View

Notes

[1] The Operation Navigator is a graphical organization aided with a tree structure that illustrates relationships between the geometry, machining method, and tool parameter groups and the operations within a program.

操作导航器是一个带有树状结构的图形组织，说明了在一个程序内几何体、加工方法、工具参数组和所有操作之间的关系。

[2] The data required to create the manufacturing plan would include information to specify the correct cutting tools and tool parameters and to specify the correct stock requirements, such as the material type, and required tolerances.

制订生产计划所需的数据包括：确定正确的切削工具及刀具参数，确定正确的加工余量需求，如材料类型，以及所需的公差。

[3] When specifying a tool for an operation, you can create a tool group by retrieving a tool from the cutting tool library (defined by the configuration file), which contains much more information about tools than is required to generate tool paths.

当操作所需的刀具确定后，用户可以通过检索刀具库（由配置文件中定义）来创建一个刀具组，其中刀具库包含的关于刀具信息比需要生成刀具路径的信息更多。

[4] The event generator, the event handler, and the definition File are dependent upon each other. Together they transform the tool path data contained in the part file into a set of formatted instructions that can be read and executed by a specific machine tool/controller combination.

事件生成器、事件处理程序和定义文件之间相互影响。他们一起将包含在部分文件中的刀具路径数据转换成一组格式化的指令，这些指令可以由一些特定的机床或控制器组合读取并运行。

New Words in Chapter 2

Feature-based Machining (FBM)			基于特征加工
Operation Navigator			操作导航器
rough	[rʌf]	vi.	粗加工
finish	['ɪnɪʃ]	vi.	精加工
assembly	[ə'semblɪ]	n.	装配，装配体
associativity	[əˌsəʊʃjə'tivəti]	n.	相关性
operation	[ˌɑpə'reʃən]	n.	操作
Machine Coordinate System (MCS)			加工坐标系
Reference Coordinate System (RCS)			参考坐标系
Work Coordinate System (WCS)			工作坐标系
parent group			父节点组
stock	[stɒk]	n.	余量
curvature	['kɜːvətʃə]	n.	曲率
blend	[blend]	n.	圆角
draft	[dræft]	n.	拔模
planar milling			平面铣
contour milling			轮廓铣
face milling			面铣削
cavity milling			型腔铣
plunge milling			插铣
Z-level Milling			等高轮廓铣
fixed-axis surface contouring			固定轴表面轮廓铣
variable-axis surface contouring			可变轴表面轮廓铣
clearance plane			安全平面
In Process Workpiece (IPW)			处理中工件
Absolute Coordinate System (ACS)			绝对坐标系
avoidance	[ə'vɒɪdəns]	n.	避让
facet	['fæsɪt]	n.	面，小平面
gouge	[gaʊdʒ]	n.	沟，圆凿
pixel	['pɪksl]	n.	像素
Cutter Location Source File (CLSF)			刀位源文件
Graphics Postprocessor Module (GPM)			图形后置处理模块

Chapter 3
UG CAM Path Settings

Objectives:
✓ To understand the path setting of CAM.
✓ To be able to specify path setting parameters.
✓ To understand the non-cutting moves.

3.1 Path settings overview

Use the Path settings options to specify parameters that help you control the tool path. You can specify cut patterns, cut levels, cutting parameters, non-cutting moves, feeds and speeds in the operation, or in the Method parent group.

(1) Specify common parameters in a method parent, so that they are inherited by all operations in the hierarchy and you do not have to repeat the selections. You can specify the following:
· Cut Method
· Tolerances.
· Stock.
· Feedrates.

The Method parent groups appear in the machining method view of the Operation Navigator.

(2) Specify operation-specific parameters in the path settings group of the operation. You can also:

· [　　▾] Select the method parent that controls inherited parameters.

· [icon] Create a new method parent for this operation.

· [icon] Edit the selections in a previously defined parent group.

(3) Specify parameters in both locations to inherit common parameters and maintain control over operation-specific parameters.

Path settings options are shared by many operations, but not necessarily all of them. The most common options are:

- Cut pattern.
- Stepover.
- Cutting parameters.
- Non-cutting moves.
- Feeds and speeds.

Path settings group options are shown in Tab.3-1.

Tab.3-1 Path Settings group options

options	Description
Method	Lets you create a new Method group or edit the selected method definition for this operation and place it in the machining method view of the Operation Navigator for use by other operations
Cut pattern	Selects the tool path pattern used to machine cut regions. The cut types available depend on the operation
Additional passes	Available for profile and standard drive cut patterns. Specifies the number of passes in addition to the pass along the boundary, so that you can remove material gradually in multiple passes
Defer cutting of regions	Lets you view just the cut regions without taking the time to view the actual machining of these regions. The software builds all the machinable regions without cutting them
Global depth per cut	The default value when a range is added. This value affects the maximum depth per cut of all cut ranges in auto generate or single mode
Cut levels	Provides greater control of cut ranges and the cut levels with each range
Cutting parameters	Modifies the cutting parameters of the operation. The available parameters are determined by the operation's type, subtype, and cut pattern
Non-cutting moves	Specifies movements that position the tool before, after, and between cutting moves. includes cutter compensation
Feeds and speeds	Specifies spindle speeds and feed rates

3.2 Cut Pattern

Cut patterns in planar and cavity milling operations determine the tool path pattern used to machine cut regions. The following remove a volume of material with parallel linear passes.
- Zig.
- Zig-Zag.
- Zig with contour.

The following remove a volume of material with a sequence of concentric cutting passes that can progress inward or outward.
- Follow periphery: follows only the periphery geometry.

· Follow part: follows all specified part geometry.

· Trochoidal: cuts using loops to limit excess stepover and control tool embedding.

The following create one or more finish passes that follow the part walls within open or closed regions.

· Profile.

· Standard drive.

User-defined or system-defined control points determine where the initial engage occurs for each cut type.

For more information on cut patterns, see Chapter 4.4 about Cut patterns of planar milling.

3.3 Stepover

Use Stepover to specify the distance between cut passes. It specifies the distance directly by entering a constant value or percentage of the tool diameter, or indirectly by entering a scallop height and allowing the system to calculate the distance between cut passes.

Three options are available for Stepover (see Fig.3-1):

(1) Constant: This allows you to specify the maximum distance between successive cut passes.

You can specify the distance in the current units or a percent of the current tool. If the specified distance between passes does not divide into the area evenly, the software reduces the distance between passes to maintain a constant stepover.

(2) Scallop: This allows you to specify the maximum height of material to leave between passes.

The software calculates the stepover distance required to leave no more than the specified scallop height between passes. The calculated stepover may vary from cut to cut depending upon the shape of the boundary. To protect the tool from a extremly heavy load when removing material, the maximum stepover distance is restricted to two thirds of the tool diameter.

(3) % of Tool flat: This allows you to specify a fixed distance among successive cut passes as a percentage of the effective tool diameter.

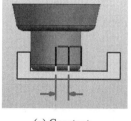　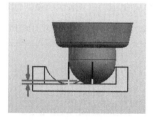　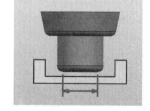

(a) Constant　　　　　　　　(b) Scallop　　　　　　　　(c) % of Tool flat

Fig.3-1　Stepover

The effective tool diameter is the diameter of the cutter that actually touches the floor of your pocket. For ball-end mills, the system uses the full tool diameter as the effective tool diameter. For all other cutters, the effective tool diameter is computed as D - 2CR. If the distance between passes does not divide into the area evenly, the software reduces the distance between passes to maintain a constant stepover.

3.4 Cut levels

Use the Cut levels command to specify cut ranges and the depth of the cut within each range. Cavity milling and Z-level milling operations complete cutting at one level before moving along the tool axis to the next level.

The following visual aids help you identify cut range and cut depth levels:

- ◁ Cut range

- ◁ Local depth per cut

Fig.3-2 shows the top level tool path. Fig.3-3 shows a lower level tool path progressing down through the cut levels.

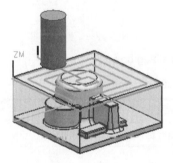 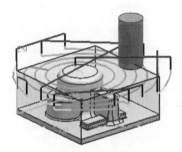

Fig.3-2 The top level tool path Fig.3-3 Lower level tool path

For more information on cut levels, see Chapter 5.4 about cut levels of cavity milling.

3.5 Cutting Parameters

3.5.1 Cutting Parameters Overview

Use the cutting parameters options to do the following.
- Define how much stock to keep on the part after cutting.

55

- Provide additional control of cut patterns, such as cut direction and cut region sequence.
- Determine the input blank and specify blank distance.
- Add and control finish passes.
- Control the cutting behavior in corners.
- Control cut order and specify how to connect cutting regions.

These options are shared by most of the processors, but not necessarily all of them. The options (see Tab.3-2) appear on several tabs in the dialog box, as shown in Fig.3-4.

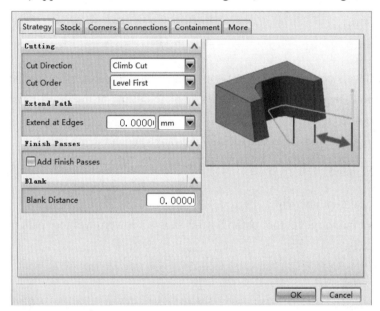

Fig.3-4 Cutting parameters dialog box

Tab.3-2 Options appear on several tabs in cutting parameters dialog box

Tab	options
Strategy	Cutting/finish pass/extend path/blank/undercuts
Stock	Stock
Corners	Path shape in corners/ feed adjustment on arcs/ feed slowdown in corners
Connections	Cut order/across voids/open passes/optimization/between levels
Containment	Blank/tool holder/small area avoidance/reference tool
More	Clearance/cut step/legacy/lower limit plane/undercuts/ramping/cleanup

3.5.2 Strategy tab

1. Cutting options

- Cutting: The available options depend on the operation type and subtype settings.
- Cut direction: Calculates the direction of cut based on boundary direction and the spindle rotation direction. The available options are shown in Tab.3-3.

Tab.3-3 The cut direction options

Options	Sketch	Description
Climb cut		Specifies that with a clockwise spindle rotation, the material is on the right side of the cutter
Conventional cut		Specifies that with a clockwise spindle rotation, the material is on the left side of the cutter
Follow boundary		Available for operations that use boundaries, and is most often used on an open boundary. It specifies the direction in which the tool travels when cutting. The tool path cuts in the direction in which you selected the boundary members
Reverse boundary		Available for operations that use boundaries, and most often used on an open boundary. It specifies the direction in which the tool travels when cutting. The tool path cuts in the direction that is opposite to direction in which you selected the boundary members
Mixed cut		Available for Z-level operations. It alternates the cut direction from level to level. In addition to climb cut and conventional cut, you also can alternate cut directions from level to level by cutting forward and backwards

· Cut order: Available for planar milling, cavity milling, and Z-level milling. It specifies how to process a cutter path with multiple regions. The available options are shown in Tab.3-4.

Tab.3-4 The cut order options

Options	Sketch	Description
Level first		Finishes each level across multiple regions before cutting to the final depth. This can be useful when working with thin wall pockets
Depth first		Cuts a single region to full depth before moving on to the next region

· Cut angle: Available for the zig, zig-zag, and zig with contour cut patterns. For face milling, cut angle is used only for the zig, and zig-zag cut patterns. It rotates the tool path with respect to the WCS. The cut angle is the orientation of the cutter path with respect to the XC-axis of the WCS at the time the angle was specified. The available options are shown in Tab.3-5.

Tab.3-5 The cut angle options

Options	Sketch	Description
Automatic		The software evaluates each cut region shape and determines an efficient cut angle that minimizes internal engage moves when cutting the region
Specify		Specifies the cut angle. The angle is measured with respect to the X-axis on the XC-YC plane of the WCS, then is projected to the floor plane
Longest line		Establishes the cut angle parallel to the longest line segment in the peripheral boundary. If the peripheral boundary does not contain a line segment, the software searches for the longest line segment in the interior boundaries
Vector		Stores the defined 3D vector as the cut direction. To define the cut angle, the software projects the 3D vector along the tool axis to the cut levels. This option enables a predictable cut angle regardless of the WCS orientation

· Pattern direction: Available for follow periphery and trochoidal cut patterns. This allows you to reverse the pocketing tool path to lessen chip interference and eliminate the need to pre-drill holes. The available options are shown in Tab.3-6.

Tab.3-6　The pattern direction options

Options	Sketch	Description
Inward		Starts cutting at the periphery of the part and finishes at the center
Outward		Starts cutting at the center of the part and finishes at the periphery

· Trochoidal settings: Available for the trochoidal cut pattern. The available options are shown in Tab.3-7.

Tab.3-7　The Trochoidal Settings Options

Options	Sketch	Description
Trochoidal width		Specifies the diameter of the trochoidal circles. The diameter is measured at the path center line
Min Trochoidal width		Specifies the smallest allowable diameter for the trochoidal circles. Use a variable width to increase your tool path control in sharp corners and narrow slots
Stepover limit		Specifies the maximum amount by which actual stepover can exceed the stepover specified in the Milling operation dialog box. Trochoidal loops prevent larger stepovers
Trochoidal Step ahead		Specifies how far apart the trochoidal circles are spaced along the tool path

· Self-intersection: Available for the standard drive cut pattern. It allows self-intersecting tool paths for each shape.

· Walls: Available for the zig, zig-zag, and follow periphery cut patterns in face milling, planar milling, and cavity milling operations. It inserts a final profile pass at each cut level to remove ridges that remain along the part walls. Wall cleanup passes are different from profile passes. The available options are shown in Tab.3-8.

Tab.3-8　The walls options

Options	Sketch	Description
Automatic		Available for the follow periphery cut pattern. Uses profile passes to remove all material without re-cutting the material
None		Does not always remove all the material, but can create a shorter tool path with fewer engages
At start/ At end		Makes an extra profile pass along the part walls to remove uncut material and re-cut some of the outer follow periphery pass
Cut walls only		Limits the cut path to walls only

2. Finish pass options

Finish passes controls the last cutting pass (or passes) the tool makes after it completes the main cutting passes. It's available for the following cut patterns in cavity milling, face milling, and planar milling:

· Follow part.
· Follow periphery.
· Trochoidal.
· Zig.
· Zig-zag.
· Zig with contour.

You can add one or more finish passes (see Fig.3-5) to the operation and request multiple finish passes with centerline cutter compensation. With contact contour cutter compensation, you can only request one finish pass.

Fig.3-5   Add Finish Passes

3. Extend path options

Extend path is available for Z-level and surface contouring operations.

(1) Extend at edges: available for flowcut, area mill, cavity milling, and Z-level profile operations (see Fig.3-6 and Fig.3-7). It requires cut area geometry, and makes the cutter go beyond the exterior edges of the cut area to machine excess material around the part. You also can use this option to add cutting moves to the start and end of tool path passes to ensure that the tool smoothly enters and exits the part.

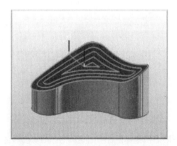   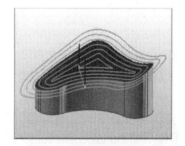

Fig.3-6   Extension of edges off   Fig.3-7   Extend at edges on

(2) Roll tool over edges: Available for contour milling and Z-level milling operations (see Fig.3-8 and Fig.3-9). It controls whether or not the tool is allowed to roll over edges when the tool

path extends beyond the edge of the part surface. When the tool path extends beyond the edge of the part surface, the tool attempts to complete the tool path while remaining in contact with the part surface. Tool rollover occurs only under the following conditions:

① When the tool path extends beyond the edge of the part surface.

② When the tool axis is independent of the part surface normal, as in fixed axis operations.

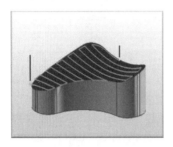 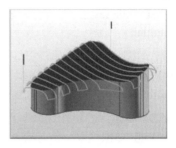

Fig.3-8　Roll tool over edges off　　　　Fig.3-9　Roll tool over edges on

(3) Continue cutting below tool contact: Continues machining the part silhouette below any reverse curvature (see Fig.3-10 and Fig.3-11).

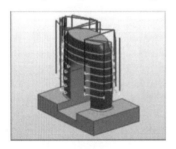 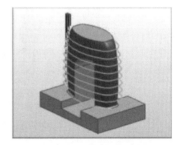

Fig.3-10　Stops machining before reverse curvature　　Fig.3-11　Continue cutting below tool contact

(4) Extend at convex corner: Available for fixed axis surface contouring operations (see in Fig.3-12 and Fig.3-13). It provides additional control over the tool path when cutting across internal convex edges to prevent the tool from dwelling on them. Performs a small lift of the tool path from the part without executing a retract/transfer/engage sequence. The lift is output as a cutting move.

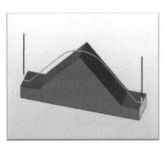 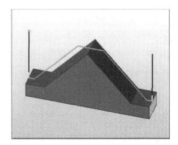

Fig.3-12　Extend at convex corner off　　　Fig.3-13　Extend at convex corner on

4. Blank options

Available for cavity milling, planar milling, and face milling operations. It specifies the offset distance applied to the part boundary or part geometry to produce the blank geometry. The specific behavior depends on the operation.

(1) In cavity milling, the blank distance (see Fig.3-14) is applied to all of the part geometry. The preferred way to specify the blank distance is to use a mill geometry group. Several cavity milling operations can then be placed in the group and share the geometry. A mill geometry group is also required for cavity milling operations to use the In Process Workpiece (IPW) option.

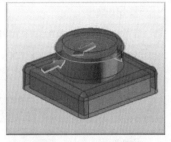

Fig.3-14 Blank distance

(2) In planar milling, the default blank distance is applied to a closed part boundary. Use blank distance instead of a blank boundary to specify a constant distance larger than the part. This is useful when working on a casting or a part that has a constant thickness of material to be removed.

(3) In face milling, each face to machine is offset along the tool axis by the blank distance value to create the blank.

5. Undercuts options

Available for face milling operations. Prevent undercutting requires more processing time to identify undercut areas. Use this option when you do not want to machine faces that lie underneath part ledges. If ignores undercut geometry, It will result in looser tolerances in processing vertical walls.

3.5.3 Stock tab

Use the options on the stock tab to specify the amount of material to remain on the part after the current operation. You can also specify the amount of material to remain after a final profile pass which will remove some or all of any specified stock.

If an operation uses boundaries, you can also specify stock requirements at the boundary level, and at the boundary member level uses custom boundary data. In milling operations, stocks applied at the custom boundary level override all other part stocks.

Use the Intol and Outtol parameters to define an allowable range (see Fig.3-15). The tool path may deviate from the actual part surfaces. Smaller values produce a smoother more accurate cut, but require more processing time.

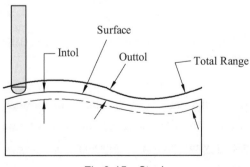

Fig.3-15 Stock

1. Stock

The stock parameters that are available for an operation are determined by the processor. The available options of stock are shown in Tab.3-9.

Tab.3-9 The stock options

Option	Sketch	Applicative processor	Description
Part stock		Face milling, planar milling, and surface contour milling	Specifies the amount of material to remain after machining
Check stock		Face milling, planar milling, cavity milling, Z-level milling, and surface contour milling	Specifies the distance to position the tool from the defined check boundary
Wall stock		Face milling area	Applies a unique stock to walls. Wall stock is applied in the plane of the cut wherever the plane of the cut intersects the wall. The tool is offset from the wall by the wall stock value plus half of the tool (D) Diameter value
Blank stock		Face milling, planar milling, and cavity, milling	Specifies the distance. The tool is offset from the defined blank geometry. Blank stock applies to blank boundaries or blank geometry that have a Tanto condition
Final floor stock		Face milling and planar milling	Sets a value for the amount of material to remain uncut. The floor stock is measured from the face plane and offset along the tool axis
Part floor stock		Cavity milling and Z-level milling	Specifies the material remaining on the floor. The stock is measured vertically along the tool axis. Part floor stock applies only to part surfaces that define cut levels, are planar, and are normal to the tool axis
Part side stock		Cavity milling and Z-level milling	Specifies the material remaining on the walls and it is measured normal to the tool axis (horizontally) at each cut level. Part Side Stock is applied to all part surfaces (planar, nonplanar, vertical, angled) from which a horizontal measurement can be taken
Use floor same as side		Cavity milling and Z-level milling	Sets Floor Stock equal to the Part Side Stock value
Use floor same as wall		Contour profile drive method in variable coutour milling	Sets Floor Stock equal to the Wall Stock value
Trim stock		Planar milling, cavity milling, and Z-level milling	Specifies the distance from the positioned tool to the defined trim boundary

2. Tolerance

Tolerance specifies how far the cutter can deviate from the part surface. Smaller Intol and Outtol values allow less deviation from the surfaces and produce smoother contours, but require more processing time because they produce more cut steps. The available options of tolerance are shown in Tab.3-10.

Tab.3-10 The tolerance options

Option	Sketch	Applicative processor	Description
Intol		Planar milling, cavity milling, Z-level milling, face milling, surface contour milling	Specifies the maximum distance by which the cutter can deviate from the intended tool path when it cuts into the part surface
Outol		Planar milling, cavity milling, Z-level milling, face milling, surface contour milling	Specifies the maximum distance which the cutter can deviate from the intended tool path when it cuts away from the part surface
Part Intol		Surface contour milling	Specifies the maximum distance which the cutter can deviate from the intended tool path when it cuts into the part surface
Part Outol		Surface contour milling	Specifies the maximum distance which the cutter can deviate from the intended tool path when it cuts away from the part surface
Boundary Intol		Surface contour milling	Specifies the maximum distance that a cutter can deviate from the intend tool path cutting inside of the boundary (violates the workpiece)
Boundary Outol		Surface contour milling	Specifies the maximum distance which the cutter can deviate from the intended tool path when it cuts into the part boundary

3.5.4 Corners

Use the options on the corners tab to smoothly transit cutting moves related to:
· Corner rounding in follow part, follow periphery, and trochoidal cut patterns.
· Stepover moves in follow part, follow periphery, and trochoidal cut patterns.
· Smoothing in the stepover move in zig and zig-zag cut patterns.

1. Path shape in corners options

(1) Convex corners: Available for planar mill and face milling operations. The available options of convex corners are shown in Tab.3-11.

Tab.3-11 The convex corners options

Options	Sketch	Description
Roll around		Transits part walls by rolling around the corner
Extend and trim		Transits part walls by extending the adjacent segments
Extend		Transits part walls by extending the adjacent segments to an intersection point

(2) Smoothing: provides options for adding arcs to the tool path. The available options of smoothing are shown in Tab.3-12.

Tab.3-12 The Smoothing Options

Options	Sketch	Description
None		Do not apply a smoothing radius to tool path corners and stepovers
All passes		Applies a smoothing radius to tool path corners and stepovers

2. Feed adjustment on arcs options

(1) None: Do not apply an adjustment in feed rate.

(2) On All Arcs: Min Compensation Factor provides the smallest slowdown factor to reduce the feed rate. Max Compensation Factor provides the largest slowdown factor to reduce the feed rate.

3. Feed slowdown in corners options

(1) None: Do not apply a slowdown in feed rate.

(2) Current tool: Uses the diameter of the current tool as a slowdown distance.

① Tool diameter percent: uses a percentage of the tool diameter as a slowdown distance.

② Slowdown percent: sets the percentage amount of slowdown from the original feed rate. The default is set to 110 percent.

③ Number of steps: sets the number of slowdown steps that are applied to the feed rate. The default is set to 1 step.

④ Minimum corner angle: sets the minimum angle that is recognized as a corner. The default is 0 degree.

⑤ Maximum corner angle: sets the maximum angle that is recognized as a corner. The default is 175 degrees.

(3) Previous tool: Uses the diameter of the previous tool as a slowdown distance.

3.5.5 Connections tab

1. Cut order options

Cut order options is available for planar milling, cavity milling, and face milling.

(1) Region sequencing provides several methods for automatically and manually specifying the order in which cut regions are machined. The available options of region sequencing are shown in Tab.3-13.

Tab.3-13 The region sequencing options

Options	Sketch	Description
Standard		Determines the order in which cut regions are machined. The software does this automatically. When machining multiple cut levels using the level first option for cut order, the processor repeats the same order for each level
Optimize		Sets the order for machining cut regions based on the most efficient machining time. The processor determines the order that requires less crossing back and forth between regions and a shorter overall distance when traversing from one region to another
Follow start points		Sets the order for machining cut regions based on the order in which the region start points were specified. These points must be active for region sequencing to use them
Follow Predrill Points		Sets the order for machining cut regions based on the order in which the Pre-Drill Engage Points were specified. It is applied the same rules as Follow start points

2. Across Voids options

Available for zig, zig-zag, and zig with Contour cut patterns in Planar Milling, and Cavity Milling, and for all cut patterns in Face Milling.

(1) Motion Type: Specifies cutter movement when there is a void. The available options of Motion Type are shown in Tab.3-14.

Tab.3-14 The Across Motion Type Options

Options	Sketch	Description
Follow		Specifies that the cutter must lift when there is a void
Cut		Specifies that the cutter maintains the cut feed rates while cutting in the same direction across a void
Traverse		Specifies that the cutter changes from a cut feed rate to a traverse feed rate when the tool is completely in the air over the void. The cutter continues to cut in the same direction

(2) Min Traverse Distance: Specifies the longest distance the software allows the tool to go through air at the cut feed rate. When Min Traverse Distance is exceeded, the feed rate changes from the cut feed rate to the traverse feed rate.

3. Open Passes options

Open Passes options is available for Follow Part cut patterns in Planar Milling and Cavity Milling. Open passes are formed when the offset passes from the part intersect with the blank of the region. The available options of Open Passes options are shown in Tab.3-15.

Tab.3-15 The open passes options

Options	Sketch	Description
Maintain cut direction		Specifies that cut direction is maintained when traversing open passes
Alternate cut direction		Alternates cut direction when traversing open passes

4. Optimization options

(1) Feed on short moves: Available for the trochoidal cut pattern. It also specifies how to connect different cutting areas within a region.

① Feed on short moves off (see Fig.3-16): Specifies that the cutter with the current transfer method to retract, and then traverse to the next location and engage.

② Feed on short moves on (see Fig.3-17): Moves the cutter along the part surface, at the stepover feed rate, until the distance is less than the max traverse distance value.

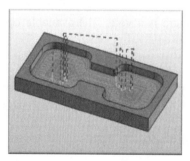 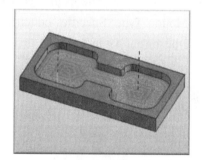

Fig.3-16 Feed on short moves off   Fig.3-17 Feed on short moves on

(2) Region connection: Available for follow periphery, follow part, and profile cut patterns in cavity milling, face milling, and planar milling.

Islands, channels, or other obstructions to the path can result in a number of subregions within the machinable region of a cut level. The subregions are connected with retract and engage moves. Region connection determines how to traverse passes and connect subregions. The processor optimizes the stepover move between passes in order to achieve a path that does not cut twice or lift the tool. The software may therefore ignore the start point of a region when the inner passes of the region are split.

① Region connection off (see Fig.3-18): Generates the tool path quickly when the part contains islands or channels between regions.

② Region connection on (see Fig.3-19): This option allows for greater prediction of where the tool path starts, and for greater feed rate control.

(3) Follow check geometry: available for planar milling and cavity milling. Lets you determine how the cutter behaves when it encounters check geometry.

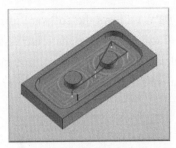
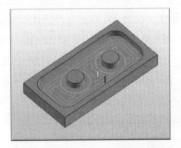

Fig.3-18 Region Connection off   Fig.3-19 Region Connection on

① Follow check geometry off (see Fig.3-20): Retracts the cutter when check geometry is recognized, and uses the specified avoidance parameters.

② Follow check geometry on (see Fig.3-21): Cuts around the identified check geometry.

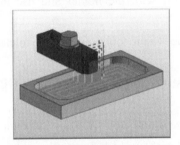
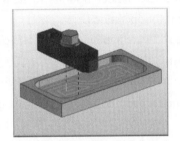

Fig.3-20 Follow check geometry off   Fig.3-21 Follow check geometry on

5. Between Levels options

Available for Z-level milling. Lets you cut all levels without lifting back up to the clearance plane. For more information, see Section 7.4 about Between levels.

3.5.6 Containment tab

1. Blank options

Blank is used to remove cutting motions that do not touch material. The finishing operations disregard the blank. Blank is not a prerequisite for multi-blade operations. However, once blank is defined it is considered for the roughing operation.

(1) Trim-by: Use trim-by to define and generate a machinable cut region. It is also available for cavity and Z-level milling. The available options of trim-by are shown in Tab.3-16.

Tab.3-16 The trim-by options

Options	Sketch	Description
None		Cuts to the existing shape of the part
Silhouette		Available for cavity milling when tolerant machining (more tab) is on. Creates blank geometry from the exterior edges (silhouette) of the selected part geometry

67

(2) In Process Workpiece: An In Process Workpiece (IPW) allows the software to recognize the material remaining from prior operations. IPW is applied to visualize the material (rest material) remaining from prior operations, to define blank material, and to check for tool collisions. Not available for plunge milling, profile cut pattern or Z-level milling operations. There are several IPW containment options:

· Use level-based (see Fig.3-22).
· Use 3D.
· Use 2D IPW.

For more information on IPW, see Section 5.2 about IPW.

2. Tool holder options

Use the tool holder options to help avoid collisions with the holder and use the shortest possible tool for operations.

· The software applies the part safe clearance distance to the tool holder shape to ensure a safe clearance from the geometry.

· The software checks for holder collisions against the In Process Workpiece (IPW) or the blank geometry, part geometry, and check geometry.

· The software removes any areas that would cause a collision from the cut regions, so the resulting tool path cuts only material that can be removed without the holder colliding.

· The software updates the removed material after each level is cut to maximize the cutting range and allow for greater holder accessibility at the lower levels.

You must cut the removed (collision) regions with a longer tool in a follow-up operation.

An effective practice is to use a sequence of cavity milling operations with progressively longer tools, tool holder checking, and the IPW. This lets you machine the most material with the shortest tool. The longer tools only machine the remaining material. To avoid tool paths that remove very little material, you can also set the minimum amount of material to remove.

Fig.3-22 Translucent level-based IPW shown with part

(1) Tool holder: available for face milling, Z-level milling, planar profile, profile 3d, solid profile 3D, and cavity milling operations, and also available for surface contouring operations like area milling and flowcut drive methods (see Fig.3-23).

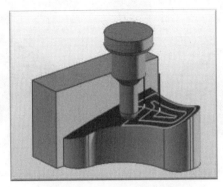

Fig.3-23 Use Tool Holder

(2) Check for IPW collisions: Available for cavity milling operations when use tool holder is selected. Lets you specify whether or not to check for IPW collisions.

· Check for IPW collisions off (see Fig.3-24): Requires less memory and improves performance if you already know how much material remains on the part. This also specify a safe clearance distance value (non-cutting moves dialog box→transfer/rapid tab) to prevent collisions.

· Check for IPW collisions on (see Fig.3-25): Prevents collisions when you do not know how much material remains on the part.

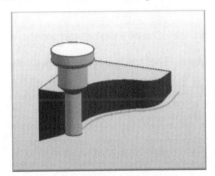 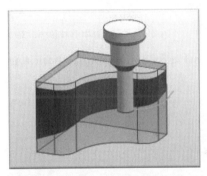

Fig.3.-24 Check for IPW collisions off   Fig.3-25 Check for IPW collisions on

(3) Suppress path if less than minimum: Lets you control whether or not to output a tool path if the operation would only remove a small amount of material.

3. Reference tool options

Use the reference tool option to eliminate tool motion where there is no material. When you create a new operation with a smaller tool that references a larger tool, the smaller tool only removes material that is not cut by the larger tool. You can place the reference tool operation before or after operations with larger tools.

(1) Place the reference tool operation before operations using the larger tool to first rough cut corners with a smaller tool.

In Fig.3-26, the yellow area shows the areas cut with the smaller tool. The tool path was contained in those areas by the reference tool. In Fig.3-27, the green area shows the areas cut with the larger tool. The tool path was previously contained in the yellow areas by the reference tool.

 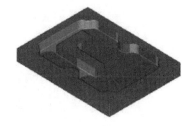

Fig.3-26 Cut with the smaller tool   Fig.3-27 Cut with the larger tool

(2) Place the reference tool operation after operations that the larger tool fail to cut areas, as shown in Fig.3-28 and Fig.3-29.

 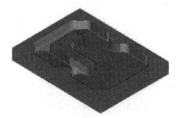

Fig.3-28 Cut with the larger tool   Fig.3-29 Cut with the smaller tool

In Fig.3-28, the part has been roughed with a large tool. The reference tool operation will remove the remaining material. In Fig.3-29, the yellow area shows the areas cut with the smaller tool. The tool path was contained in the yellow areas by the reference tool.

3.6 Non-cutting Moves

3.6.1 Non-cutting moves overview

Use the Non-cutting moves (see Fig.3-30) command to avoid collisions with the part or fixture devices. Non-cutting moves work as the following:

· Position the tool before, after, and between cut moves.

· Create non-cutting tool path segments that are connected with cut move segments to form a complete tool path within a single operation.

· Non-cutting moves in an operation can be as simple as a single engage and retract, or as comprehensive as a series of custom engage, retract and transfer (approach, traverse, departure) moves designed to accommodate multiple part surfaces, check surfaces, and lifts between cut passes.

· Non-cutting moves include cutter compensation because cutter compensation is activated during a non cutting move.

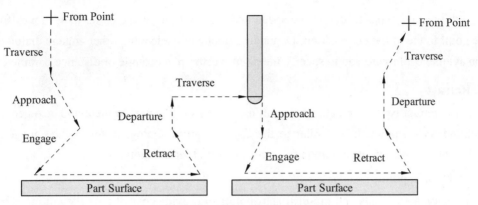

Fig.3-30   Non-cutting moves

Gouge-free non-cutting tool paths are accomplished with collision check and the precise control you have over the construction of the non-cutting tool path.

To facilitate precise tool control, all non-cutting tool movements are internally calculated forward (in the direction of tool movement), except for engage and approach, which are constructed backward from the part surface to ensure the spatial relationship with the part prior to cutting (see Fig.3-31).

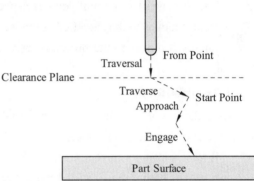

Fig.3-31   Forward construction of traverse moves

3.6.2   Non-cutting move types

Non-cutting move types (see Tab.3-17) are determined by the operation type and subtype. The moves are organized by function, and are placed on property tabs in the dialog box.

Tab.3-17   Non-cutting move types

Move Type	Description
Engage	Specifies the move that takes the tool to the start of the cutting pass
Retract	Specifies the move that takes the tool away from the end of the cutting pass
Start/Drill points	Specifies the tool engage position and the stepover direction by identifying a user defined or default region start point. The pre-drill points option specifies engage locations within previously drilled holes or other vacancies in the blank material
Transfer/Rapid	Specifies how to move from one cutting pass to another
Avoidance	Allows you to specify, activate, cancel, and manipulate points, lines, or symbols. These points, lines or symbols help you define a tool clearance motion before and after the cut motion

1. Engage

Use the engage types to define the speed and tool motion for the cutter as it moves from the engage point to the initial cut position. Depending upon your selection, other engage dialog options become available allowing you to specify the point, vector, plane, angle or distance parameters.

2. Retract

Use the retract types to create a non-cutting move from the part to the avoidance geometry or to a defined retract point. It is similar to the engage motion dialog. Refer to the engage motion dialog section for the options common to both engage and retract motions.

3. Start/Drill points

Use region start points and pre-drill start points to provide control over the cutting start point within single and multiple cut regions.

When you use these options, the tool does the following:

· First it moves to the point you have specified.

· Next it moves to the specified cut level.

· Finally it moves to the processor generated start point and then generates the remainder of the path.

4. Transfer/Rapid

Uses the options on the transfer/rapid tab to specify how to move from one cutting pass to another. Generally, the cutter makes the following moves (see Fig.3-32).

· It moves from its current position to the specified plane.

· It moves within the specified plane to a position above the start of the engage move. If no engage move is specified, it moves above the cut point.

· It moves from the specified plane to the start of the engage move. If no engage move is specified, it moves above the cut point.

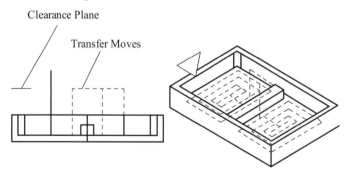

Fig.3-32　Transfer/Rapid moves

If the clearance plane is not perpendicular to the tool axis, the transfer move depends on the angle between the clearance plane normal and the tool axis.

If the angle between the tool axis and clearance plane normal is less than or equal to 45 degrees, the transfer move is along the tool axis, as shown in Fig.3-33 (a). If the angle between the tool axis and clearance plane normal is greater than 45 degrees, the transfer move is perpendicular to the tool axis, as shown in Fig.3-33 (b).

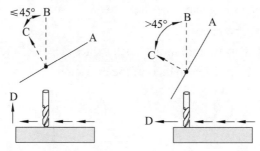

A = clearance plane, B = tool axis, C = clearance plane normal, D = transfer move
(a) the angle less than or equal to 45°   (b) the angle greater than 45°

Fig.3-33   The transfer movement when the clearance plane is not perpendicular

Use the non-cutting moves dialog box to set the non-cutting moves options. The options are slightly different between different processors, as shown in Fig.3-34.

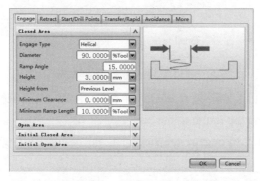

Fig.3-34   The non-cutting moves dialog box

3.6.3   Open and closed areas

Open areas (see Fig.3-35) are regions where the tool can reach the current cut level, before removing material. Closed areas are regions where the tool must start removing material before reaching the current cut level.

Open and closed areas are determined by the following:
· Geometry.
· Operation.
· Cut pattern.
· Trim boundaries.

If you use trim boundaries to localize the cut area, NX assumes a closed area even if there is only a little bit of blank outside the trim boundary, or the trim boundary and blank are coincident.

The system recognizes the following regions as open areas:
· Any pre-drilled hole indicated by user defined points.

· Any position that is opposite the material side of a boundary segment whose tool position is on.

· Any position along an imaginary line that extends beyond the start point or end point of a tool path that follows an open boundary.

· The inside of a pocket away from initial stock values (see the following note).

· A region already cut by previous passes of the operation.

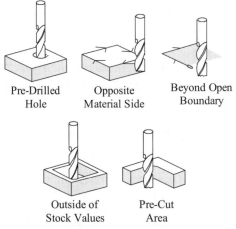

Pre-Drilled Hole   Opposite Material Side   Beyond Open Boundary

Outside of Stock Values   Pre-Cut Area

Fig.3-35   Open Areas

If you specify part stock values that are too thick, you can effectively close up a pocket, requiring the system to ramp the tool into the part.

Before the tool is ready to engage the side material, it goes through the following steps (see Fig.3-36):

(1) The engage starts at the specified clearance plane or just above either the blank plane or the previous cut plane.

(2) The tool is first lowered into the open area just above the cutting plane using an approach feed rate.

(3) The tool is then lowered to the cutting plane using an engage feed rate and, is now ready to engage the side material.

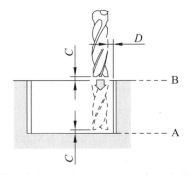

A = cut level plane, B = blank plane, C = minimum clearance (closed area), D = minimum clearance (open area)

Fig.3-36   Approach move to a point in an open area

3.6.4 Non-cutting moves options

The non-cutting moves options are slightly different between planar operations and surface contouring operations. The following options are about planar operations, which can be referred to the surface contouring operations.

1. Engage tab

Engage type: Let you define the speed and tool motion for the cutter as it moves from the engage point to the initial cut position. The available options of engage type for closed area are shown in Tab.3-18.

Tab.3-18 The engage type options for closed area

Options	Sketch	Description
Same as open area	—	Treats closed areas like open areas, and uses the Open Area move definitions
Helical		Creates a collision-free, helix-shaped engage move at the first cutting motion. The part and check geometry is avoided by the minimum clearance distance. The size of the helix varies from the requested size down to the minimum size that is allowed
Ramp on shape		Creates a ramping engage move that follows the shape of the first cutting motions. If the minimum clearance value is greater than 0, the shape can be modified by the part or check offset profile
Plunge		Engages directly into the part from the specified height. to avoid a collision, the height value must be greater than the material on the face
None		Do not output any engage moves. the software eliminates the corresponding approach move at the beginning of the tool path, and eliminates the departure move at the end of the tool path

The available options of engage type for open area are shown in Tab.3-19. If the engage move is inside the minimum clearance offset value, the move is extended to ensure the engage is the minimum clearance distance away from the part geometry.

Tab.3-19 The engage type options for open area

Options	Sketch	Description
Same as closed area	—	No open area moves are attempted and the closed area default is used
Linear		Creates an engage move at the specified distance in the same direction as the first cut motion
Linear-relative to cut		Creates a linear engage move that is tangent (if possible) to the tool path. This behaves the same as the linear option, except the swing angle is always relative to the cut direction
Arc		Creates a circular engage move tangent (if possible) to the start of the cutting move. The arc angle and the arc radius determine where the circular move starts. If necessary, a linear move is added to start the engage at the specified minimum clearance distance away from the part

Continue

Options	Sketch	Description
Point		Specifies a point for linear engages to start. Add a radius (A) to smoothly transit from the linear engage move to the cutting move on part material
Linear — along vector		Specifies the engage direction. Use the vector constructor to define the engage direction
Angle plane		Specifies a plane to start from. Swing angle and ramp angle define the engage direction. The plane defines the length
Vector plane		Specifies a plane to start from. Use the vector constructor to define the engage direction. The plane defines the length
None		Does not create an engage move. The approach move (if required) connects directly to the cut move

2. Retract tab

Retract type: Lets you create a non-cutting move from the part to the avoidance geometry or to a defined retract point. It is similar to the engage type. Refer to the engage type options.

3. Start/Drill points tab

(1) Overlap distance: Specifies the total overlap distance between the engage and retract moves (see Fig.3-37). This option ensures a full cleanup at the point where engage and retract moves occur. The paths overlap equally on each side of the original start point of the cutting path.

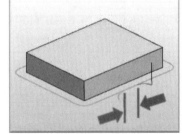

Fig.3-37　Overlap distance

(2) Region start points: Specifies where to start machining. The available options of default region start are shown in Tab.3-20.

Tab.3-20　The default region start options

Options	Sketch	Description
Mid point		Starts the tool path at the midpoint of the longest linear edge that falls within the cut region. If there are no linear edges, the longest length segment is used
Corner		Starts at the start point of the boundary specification

(3) Pre-drill points: Represent pre-drilled holes for the cutter to drop into and start machining

without any special engage.

4. Transfer/Rapid tab

(1) Clearance: Let you specify how to move from one cutting pass to another. The available options of clearance are shown in Tab.3-21.

Tab.3-21 The clearance options

Options	Sketch	Description
Use inherited		Use the clearance plane specified in the MCS
None		Do not use a clearance plane
Automatic plane		Add the safe clearance distance value to a plane that just clears the part geometry
Plane		Specifies the clearance plane for this operation, using the plane constructor to define a clearance plane

(2) Between regions: Controls retracts, transfers, and engages added to clear obstructions between different cut regions. The available options of transfer type are shown in Tab.3-22.

Tab.3-22 The transfer type options

Options	Sketch	Description
Clearance - shortest distance		All moves return to a identified clearance plane based on the shortest distance
Clearance - cut plane		All moves return to the clearance geometry along the cutting plane
Previous plane		All moves return to a previous cut level where the tool transfer can be made safely so it can move along the plane to a new cut region
Direct		Makes a direct connection transferring between the two positions
Lowest safe Z		Applies a direct move first. If the move is not gouge free, a previous safe Z-level plane is used
Blank plane		Causes the tool to transfer along the plane defined by the upper level of the material to be removed

(3) Within Regions: Controls the retract, transfer, and engage moves added to clear material

within a cut region, or between levels of a cut feature. The available options of Transfer using are shown in Tab.3-23.

Tab.3-23 The transfer using options

Options	Sketch	Description
Engage/retract		Uses the default engage/retract definition
Lift and plunge		Produces engages and retracts with vertical moves. Enter a lift/plunge height

(4) Initial and final: Controls the initial move of the operation to the first cut region or first cut level, and the final move of the operation away from the last cutting location. The available options of approach type and departure type are shown in Tab.3-24.

Tab.3-24 The approach type and departure type options

Options	Sketch	Description
Clearance - tool axis		Create the approach move along the tool axis direction. From an identified clearance plane prior to the engage move
Clearance - shortest distance		Create the approach move based on the shortest distance. From an identified clearance plane prior to the engage move
Clearance - cut plane		Create the approach move based on a cutting plane. From an identified clearance plane prior to the engage move
Relative plane		Define a plane above the initial engage point. The approach moves from this plane to the initial engage point. The distance of the plane along the tool axis is set by the safe clearance distance value in the within regions group
Blank plane		Create an approach move along the plane defined by the upper level of the material to be removed
None		Do not add an initial approach move

5. Avoidance tab

Avoidance tab options allows you to specify, activate, cancel, and manipulate points, lines, or symbols. These points, lines or symbols help you define a tool clearance motion before and after the cut motion.

3.7 Feeds and speeds

3.7.1 Feeds and speeds overview

Use the feeds and speeds command to create several types of tool motion during the progress of your tool path.

You can specify feed rates within an operation or a method group, in one of the following ways:
- Inches per minute (IPM).
- Inches per revolution (IPR).
- Millimeter per minute (MMPM).
- Millimeter per revolution (MMPR) for metric parts.

You can specify spindle speed and rotation direction within an operation in one of the following ways:
- Clockwise (CLW).
- Counter-Clockwise (CCLW).
- Revolutions per minute (RPM).
- Surface feet per minute (SFM).
- Surface millimeters per minute (SMM).

3.7.2 Feeds and speeds dialog box options

You can set feed rates and spindle speeds when you create or edit a milling operation or a method parent. The feeds and speeds dialog box is shown in Fig.3-38.

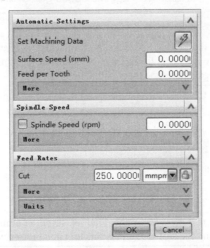

Fig.3-38 The feeds and speeds dialog box

1. Automatic settings

Use set machining data to calculate parameter settings, including depth of cut, stepover,

79

spindle speeds and feed rates.

The values are based on the following parameters you specify:

· Part material.

· Tool material.

· Cut method.

· Depth of cut.

The parameters match machining data library values. Set machining data recognizes the parameters and calculates the feeds and speeds for the selected operation.

When you click the set machining data option in a cutting operation, the feed rates, spindle speeds, cut depths, and cut levels are updated using the data you selected. Automatic Settings group is shown in Tab.3-25.

Tab.3-25 Automatic settings group

Item	Description
Set machining data	Sets feed rates, spindle speeds, cut depths and step over automatically
Surface speed (sfm)	Specifies the cutting speed of the tool, which is measured at the cutting edge of each tooth in surface feet or meters per minute. Surface speed is used to calculate spindle speed and feed per tooth. Changing this value recalculates the other parameters
Feed per tooth	Measures the amount of material removed per tooth in inches or millimeters. This value is used to calculate the cut feed rate. Changing this value recalculates the cut feed rate

2. Spindle speed group

Spindle speed group is shown in Tab.3-26.

Tab.3-26 Spindle speed group

Item	Description
Spindle speed (RPM)	Specifies the cutting speed of the tool which is measured in revolutions per minute. The software uses spindle speed to calculate surface speed and feed per tooth. Changing this value recalculates the other parameters
Output mode	· RPM — Defines the spindle speed to be in revolutions per minute. · SFM — Defines the spindle speed to be in surface feet per minute. · SMM — Defines the spindle speed to be in surface meters per minute
Range status	· Low = 1 · Medium = 2 · High = 2 if there are only 2 ranges. · High = 3 if there are more than 2 ranges
Text status	Specifies text to add to the load or turret command during CLS output. During postprocessing the software stores this text in a mom variable

3. Feed rates group

Feed rates group is shown in Tab.3-27.

Tab.3-27 Feed rates group

Item	Description
Cut	The feed rate given for the tool movement while the cutter is in contact with part geometry
Rapid	Applies only to the next GOTO point in the tool path and CLSF. Subsequent moves with the last specified feed rate
Approach	The feed rate given for the tool movement from the start point to the engage position. In planar and cavity milling operations that use multiple levels, the Approach feed rate is used to control the feed from one level to the next
Engage	The feed rate given for the tool movement from the engage position to the initial cutting position. This feed rate also applies to the returning feed rate when the tool returns to the workpiece after lifts
First cut	A cut move that is embedded in the material to cut. It violates the specified stepover by the stepover limit.
Stepover	The feed rate for the cutter as it moves to the next parallel pass. This does not apply if the tool lifts from the work surface. Therefore, the stepover feed rate applies only to modules allowing zig-zag tool paths
Departure	The feed rate given for the tool movement from the retract move, or the traversal or the return move
Traversal	The feed rate used for fast horizontal non-cutting tool motion when the transfer method option on the engage/retract menu has a status of previous level
Return	The feed rate for the tool move to the return point
Retract	The feed rate given for the tool movement to the retract position from the final tool path cutting position

If engage, first cut, or stepover is set to a feed rate of zero, the move is output at the same feed rate value that the cut feed rate is set to. (Cut can not be set to zero.)

Any motion type (except circular interpolation) with an output feed rate value of zero will typically result in G00 motion from the post.

If feed rate is not set for rapid, traversal, approach or departure, these moves would follow G00 (shown in dashed lines.) Exception: If rapid has a value entered, then neither rapid nor traversal moves are G00. Traversal would use the modal feed rate.

3.8 Machine control

3.8.1 Machine control overview

Use the machine control options to add information to your tool path. You can:
· Create or edit start of path events and end of path events.
· Add motion output settings to your existing tool paths.

Machine control events provide additional control for movement of the tool with respect to the part. These events are interpreted by the NX postprocessor or sent to the CLSF file if one is generated.

3.8.2 Create or edit start and end of path events

Start of path events and end of path events specify machine control events such as tool changes, coolant on or off, user-defined start and end events, or special post commands. They are used to generate machine code, usually for auxiliary functions. Post processors are used to interpret the machine code for output. The postprocessor determines how the events are finally output to the machine file or CLSF file.

You can define Machine Control options at the following levels:
· Boundaries.
· Boundary members. These are identified using custom boundary data.
· Boundary groups.

You can create or edit start of path events and end of path events in operations.

(1) In the operation dialog box, the machine control group, click Edit next to start of path events or end of path events.

(2) In the User defined events dialog box, click an event from the available list and then click add.

(3) In the Event dialog box, select the parameters for the event and click OK. The event is added to the defined list in the User defined events dialog box.

(4) Click the event in the defined list and then click Edit.

(5) Update the parameters as desired.

(6) In the Event dialog box, click OK to save the event changes.

(7) In the User defined events dialog box, click OK to save all the user defined events.

Start and end of path events in UG NX are shown in Tab.3-28.

Tab.3-28 Start and end of path events (partially)

Events	Description
Auxfun	Lets you output an auxiliary function code (M code) to the machine tool controller, and corresponds to the AUXFUN/ command. You can enter any number from 0 to 99 in the Auxfun Value box
Clamp	Lets you clamp an axis when the axis is not moving
Coolant On	Lets you add coolant events
Coolant Off	Cancels coolant events
Cutter Compensation	Lets you add cutter compensation events. Use this event only if cutter compensation in non-cutting moves will not work for your application
Dwell	Lets you add dwell events

3.8.3 Motion output

Many machine controllers allow you to move the tool along a true circular path or along a rational B-spline curve (NURBS). Motion Output allows you to choose whether potential circular tool motions or NURBS that will be included in your tool path. When you use this option, the system will convert a series of linear moves to a single circular move or convert linear/circular moves to a rational B-spline curve if enough of the GOTO points lie on the same arc.

Circle records output to the CLSF are generated from either selected geometry or the processor. If the radius of the circle is outside the range of .0001 - 999.9999 inches or .001 - 9999.999 millimeters, then the current postprocessor will output a warning indicating that the radius is out of range. This, however, does not indicate an error since the postprocessor will automatically convert the arc into one or more linear segments.

The type of motion output in UG NX are shown in Tab.3-29.

Tab.3-29 The type of motion output

Types	Description
Linear only	Available for milling and turning operations. Converts any circular tool movements to a series of linear GOTO moves. No circle records are generated
Circular - perp to TA	Available for milling operations. Generates all of the possible circular tool motions that lie in planes that are normal to the tool axis
Circular - Perp/par to TA	Available for milling operations. Generates all of the possible circular tool motions that lie in planes that are normal or parallel to the tool axis
Nurbs	Available for fixed-axis surface contouring, planar milling, cavity milling, face milling. Lets you control the surface finish by determining how accurately the tool path follows non-uniform rational B-spline curves (Nurbs)
Circular	Generates all possible circular tool motions
Sinumerik spline	Outputs splines for Siemens Sinumerik controllers. The Sinumerik spline output is optimized for zig cut patterns in finishing operations, and is best suited for the area mill drive method

 Notes

[1] If an operation uses boundaries, you can also specify stock requirements at the boundary level, and at the boundary member level you can use, custom boundary data.

如果在操作中使用了边界，你也可以指定在边界层的余量需求以及在边界段层中使用自定义边界数据。

[2] An In Process Workpiece (IPW) allows the software to recognize the material remaining from prior operations. IPW is used to visualize the material (rest material) remaining from prior operations, define blank material, and check for tool collisions.

处理中工件允许软件识别先前操作的剩余材料。使用处理中工件可以显示先前操作剩余量，定义毛坯材料，并检查工具的碰撞。

[3] To facilitate precise tool control, all non-cutting tool movements are internally calculated forward (in the direction of tool movement), except for engage and approach, which are constructed backward from the part surface to ensure the spatial relationship with the part prior to cutting.

为了便于精确的刀具控制，在刀具移动方向上，提前在内部计算了所有的非切削刀具运动，但这不包括进刀和逼近，它们是从零件表面向后构造，以确保与之前切割零件的空间关系。

[4] Non-cutting moves in an operation can be as simple as a single engage and retract, or as comprehensive as a series of custom engage, retract and transfer (approach, traverse, departure) moves designed to accommodate multiple part surfaces, check surfaces, and lifts between cut passes.

在操作中，非切削运动如同单一的进刀和退刀般简单，或者如同一系列自定义进刀，退刀，转移（接近，横穿，离开）运动般综合，旨在切削通路中适应多种零件表面、检查面，以及升降设备。

New Words in Chapter 3

pass	[pɑːs]	n.	刀路
zig	[zɪg]	n.	单向切削
zig-zag	['zɪgzæg]	n.	之字形，往复式切削
zig with contour			单项轮廓切削
follow periphery			跟随周边
follow part			跟随工件
trochoidal	[trəʊ'kɒidəl]	adj.	摆线的，摆线切削
profile		n.	轮廓，轮廓切削
stepover	['step,əʊvə]	n.	步距
scallop	['skæləp]	n.	残余高度
climb cut			顺铣
conventional cut			逆铣
convex	['kɒnveks]	n.	凸面，凸状
undercut	['ʌndəkʌt]	vi.	底切
silhouette	[,sɪlu'et]	n.	轮廓
flow cut			清根
engage	[ɪn'gedʒ]	vi.	进刀
retract	[rɪ'trækt]	vi.	退刀
traverse	[trə'vɜːrs]	vi.	横越
perpendicular	['pɜːp(ə)n'dɪkjʊlə]	adj.	垂直的，正交的
helical	['helɪkl]	adj.	螺旋形的
clockwise	['klɒkwaɪz]	adj.	顺时针的
counter clockwise			逆时针的
Non-Uniform Rational B-Spline (NURBS)			非均匀有理B样条

Chapter 4
Planar Milling

Objectives:
✓ To understand the planar milling.
✓ To be able to create a planar milling.

4.1 Planar milling & Path Settings overview

4.1.1 What's planar milliing

Planar milling is used to create tool paths along vertical walls or walls that are parallel to the tool axis. Boundaries are used to contain the tool path, the tool path may be single pass, multiple passes or pocket. The cut volume may be enclosed by pocket or by Blank boundaries surrounding the part boundaries.

In Fig.4-1, a blank boundary defines the material to be removed and the part boundary defines the finished part. The floor plane defines the final depth of the tool path. Check and trim boundaries can also be used to further contain the tool path. Boundaries can be selected from faces, curves, edges or points, which are associative to the selected geometry.

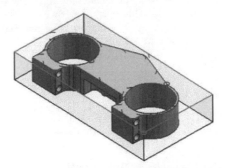

Fig.4-1   Planar milling

Planar milling operations contribute to the level-based IPW for a cavity milling operation. In planar milling:

· You create boundaries from faces, edges, curves and points to contain the tool path.

- You select the floor plane as the final depth of the tool path.
- You can select cut levels using different methods that are unique to planar milling.
- You can remove material as a cut volume using pocketing methods or by creating profile that cuts along the part boundaries.

4.1.2 Planar milling operation subtypes

Operation subtypes that use the planar milling processor (see Tab.4-1) follow 2D boundaries to remove material along vertical walls, or walls that are parallel to the tool axis. Planar milling removes material in planar levels that are perpendicular to a fixed tool axis. The parts to machine include planar islands and planar floors that are normal to the tool axis.

Tab.4-1 Planar milling operation subtypes

Icon	Subtype	Description
	Planar mill	The main Planar milling operation subtype
	Rough follow	A Planar milling operation subtype that is customized with a follow part cut pattern
	Rough zigzag	A planar milling operation subtype that is customized with a zigzag cut pattern
	Rough zig	A planar milling operation subtype that is customized with a zig with Contour cut pattern
	Cleanup corners	A planar milling operation subtype that is customized with a follow part cut pattern and uses a 2D IPW from the previous operation. Rough zig is often used to clean out corners where material is left by the previous tool
	Finish Walls	A planar milling operation subtype that is customized with a profile cut pattern. This operation leaves the inherited part stock on the walls, and has the final floor stock set to 0.25 mm
	Finish Floor	A planar milling operation subtype that is customized with a follow part cut pattern to finish the floor only, and leave stock on the walls. This operation leaves no final floor stock, but has the part stock set 0.12 mm to leave material on the walls. Part stock is not inherited in this operation

4.2 Manufacturing boundaries

4.2.1 Manufacturing boundaries overview

Use boundaries in manufacturing to define the regions to machine. Part, blank, and check boundaries are mainly used for planar milling. Newer processors, such as cavity milling or surface contouring, use part geometry instead of boundaries.

All manufacturing boundaries (see Fig.4-2) have the following characteristics.

(1) They define a path or area.

A single boundary (such as a pocket or closed area) contains the tool. Combined boundaries (such as a pocket with an island) can both contain and exclude the tool.

(2) They consist of members.

Each member is a segment (edge of a face or a curve) containing attributes such as Intol/Outtol, side stock, and tool position.

Individual members are easily identified by their tool position indicators.

(3) They include a start point, tool position, direction, material side, and type.

(4) They are always planar.

(5) They are displayed as temporary entities and disappear whenever the screen is refreshed.

(6) They have a start point and forward direction.

· When you create a boundary from a face, the software automatically establishes the boundary direction.

· When you create a boundary from curves, edges, or points, you establish the boundary direction. The direction is from the first item you select to the second item you select.

After the boundary direction has been established, you cannot edit it or reverse it.

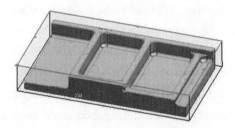

Fig.4-2　The manufacturing boundary

4.2.2　Permanent boundaries

Before mill boundary groups existed, permanent boundaries were used to create boundaries to share among multiple operations. The functionality still exists in the software for legacy parts. However, manufacturing boundaries and boundary geometry groups are the recommended method to share boundaries. Permanent boundaries can only be created by selecting curves and edges. Once created, they cannot be edited or moved.

Permanent boundaries continually display on the screen like any other permanent geometric entity (line, arc, etc.) and can be used in any machining module requiring boundaries.

Manufacturing boundaries have many advantages over permanent boundaries.

· They can be created from curves, edges, existing permanent boundaries, planar faces, and points.

· They can be edited.

· They can be customized with Intol/Outtol values, stock, and cut feed rates.

· They can easily be used to create permanent boundaries.

· They are associative to solid edges.

4.2.3 Boundaries associativity

Boundaries are associative to the solid geometry from which they were created. When the parent geometry changes, the boundary is updated to reflect those changes as shown in Fig.4-3.

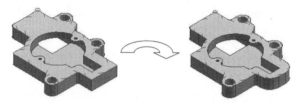

Fig.4-3 Updated boundary

(1) Manufacturing boundaries created from faces.

Boundaries are associative to the solid face and the holes (or islands) in that face. If you modify the solid face, the corresponding boundaries are updated. If you replace a boundary, or append a new boundary, associativity is affected in the following ways:

· The associativity between the boundaries and the solid face is removed.

· The associativity between the original boundaries and the edges of the face is maintained. Changing the face automatically updates the face boundaries.

(2) Manufacturing boundaries created from permanent boundaries.

When you create a temporary manufacturing boundary from a permanent boundary, the temporary boundary is associative to the curves and edges from which the permanent boundary was created. Because the temporary boundary is not associative to the permanent boundary, you can delete the permanent boundary without affecting the temporary boundary.

(3) Manufacturing boundaries created from curves.

When you create a temporary manufacturing boundary from curves, the temporary boundary is associative to the curves. Editing curves in the modeling application affects the associated manufacturing boundaries.

(4) Non-associative boundaries.

Manufacturing boundaries are non-associative only if you create them with points and clear the associative check box in the Point dialog box.

4.2.4 Create boundary options

You can create new boundaries from existing curves and edges, as shown in Fig.4-4. You may enter the name of the object you wish to select or select the curves and edges to define the boundary.

Fig.4-4　Create boundary

The options of Create boundary dialog box is shown in Tab.4-2.

Tab.4-2　Options of Create boundary dialog box

Option	Description
Chaining	Chaining provides a fast method for selecting multiple curves. Select the beginning curve away from the end that is expected to be the start point of the boundary, then select the end of the boundary. All joining curves between the two curves will also be selected
Class selection	Lets you select object by class
Ignore holes	Do not extend boundaries to include holes
Ignore islands	Do not extend boundaries to include islands
Ignore chamfers	Do not extend boundaries to include chamfers
Boundary plane	Lets you use the plane subfunction to define the plane onto which the selected geometry will be projected and the boundary will be created
Boundary type	Lets you define the boundary as open or closed
Material side	Lets you define the side of the boundary where you do not want to cut

1. Ignore Chamfers limitations

The ignore chamfers option does not extend boundaries to include chamfers if any of the following conditions apply (see Fig.4-5).

· The selected face is not planar.
· There are multiple-level chamfers.
· The chamfers have holes or islands.
· Faces adjacent to the chamfers are not perpendicular to the selected planar face.

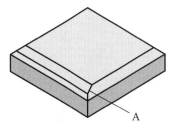 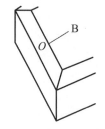 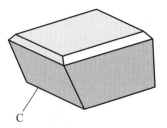

(a) Multiple level chamfer   (b) Chamfer with a hole   (c) Chamfer with adjacent non-perpendicular face

Fig.4-5 Ignore Chamfers

2. Open and closed boundaries

An open boundary (see Fig.4-6) has the following characteristics.

· It defines a path.

· It has left and right material sides that are determined by the boundary direction.

A closed boundary (see Fig.4-7) has the following characteristics.

· It defines an area. A common point defines the start of the first segment and the end of the last segment.

· It has inside and outside material sides. The boundary direction is not relevant for closed boundaries.

Fig.4-6 Open boundary   Fig.4-7 Closed boundaries

You can change the boundary type from open to closed or vice versa, if it was created from curves or edges. The software always assumes a boundary created from points is closed.

· If the first curve of an open boundary is closed, the boundary starts at the end point which is nearest the cursor position and follows along the curve towards the intersection with the next curve.

· If the last curve of an open boundary is closed, the boundary follows along the curve from the intersection with the previous curve toward the cursor position, and ends at the endpoint of the closed curve.

· If the curves or edges you select do not form a closed area, the software extends them until it is possible to form a closed boundary.

3. Specifying the material side

The material side is the side of the boundary where you do not want to cut. You can use trim boundaries to prevent the tool path from being generated on the material side (see Fig.4-8).

For part, blank, check, and drive boundaries,

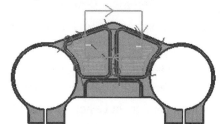

Fig.4-8 Trim boundary with material side inside

specify the side on which the material is located. The material side designation is different for open and closed boundaries.

· For open boundaries, specify the right or left side of the boundary as shown in Fig.4-9.
· For closed boundaries, specify the inside or outside of the boundary as shown in Fig.4-10.

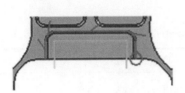
Fig.4-9　Material side of open boundary

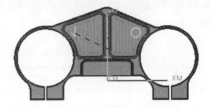
Fig.4-10　Material side of closed boundaries

4.2.5　Create boundary

This example shows how to create a boundary from a face within a mill Bnd parent group.

(1) On the insert toolbar, click create geometry or choose insert→geometry.

(2) In the Create geometry dialog box, click mill Bnd. Then a mill Bnd geometry parent is created, and is visible in the geometry view of the Operation Navigator.

(3) In the mill Bnd dialog box, click specify part boundaries.

(4) In the Part boundaries dialog box do the following:
① Click the Main tab.
② Under filter type, click Face Boundary.
③ Select the ignore holes, ignore islands check boxes and clear ignore chamfers check boxes.
④ Under material side, select inside to retain the material inside of the closed boundary.
⑤ Select the face or faces that will define the boundary as shown in Fig.4-11.
⑥ Click OK to close the Part boundaries dialog box and save the boundary.

(5) In the Mill bnd dialog box, click Display to view the boundary (see Fig.4-12).

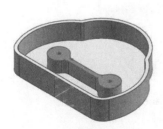
Fig.4-11　Select face boundary

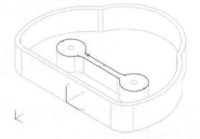
Fig.4-12　View the boundaries

4.3　Planar milling geometry

4.3.1　Valid geometry for planar milling

The following types (shown as Tab.4-3) of geometry can be specified for planar milling operations.

Tab.4-3　Valid geometry for face milling

Icon	Geometry	Description
	Part boundaries	Define the geometry to machine
	Blank boundaries	Specify the material to cut from
	Check boundaries	In addition to the specified part geometry, check boundaries are used to define areas the tool should avoid
	Trim boundaries	Limit the cut regions at each cut level
	Floor	Defines the last cut level in the operation

1. Part boundaries

Part boundaries are used to define the geometry to machine. Fig.4-13 shows a part boundary that contains the tool used to cut the part in a planar milling operation. The floor defines the depth.

2. Blank boundaries

Blank Boundaries are used to specify the material to cut from. You can use part and blank boundaries together to define the cut volume. Fig.4-14 shows a part where the difference in volumes between a single blank boundary and multiple part boundaries defines the cut volume.

Fig.4-13　Part boundary

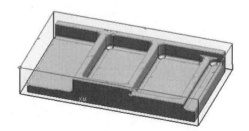

Fig.4-14　Cut volume defined by part and blank boundaries

3. Check boundaries

Check boundaries, in addition to the specified part geometry, are used to define areas the tool should avoid. You can define the part geometry by selecting boundaries or by another method. You can apply a check stock value to check boundaries.

Check boundaries always have a Tanto tool position. The normal of a check boundary plane must be parallel with the tool axis. Fig.4-15 shows check boundaries used to define clamps for a planar milling operation.

4. Trim boundaries

Trim boundaries are used to limit the cut regions at each cut level. Combine trim boundaries with the specified part geometry to discard the cut region outside of the trim boundary. For example, you can define trim boundaries so that an operation only cuts the areas where a previous operation left material underneath clamps (see Fig.4-16).

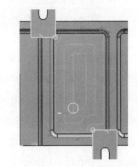

Fig.4-15 Check boundaries

Valid selection options include:
- Faces.
- Curves.
- Edges.
- Points.
- Permanent boundaries.

Fig.4-16 Trim boundary, material retained outside

The software identifies areas where trim boundaries overlap the specified part geometry, then discards the cut region outside the trim boundary. Trim boundaries always have an on tool position.

Do not use trim boundaries where the tool cannot safely engage into the cut region. For example, if you use a trim boundary to cut a small area inside the blank, the engage moves to the trimmed area may start inside of the blank. If the blank material was not removed by a previous operation, the tool will collide with the blank material.

5. Floor planes

The floor plane is the floor geometry that you select with the specify floor option in a planar milling operation. The floor plane defines the last cut level in the operation. All cut levels are generated parallel to the floor plane. You can specify only one floor plane per operation. If you select new floor geometry, the existing floor plane is replaced.

In Fig.4-17, the floor plane is selected inside the pocket. In Fig.4-18, the floor plane is selected as the bottom of the part.

 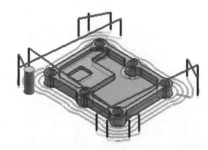

Fig.4-17 Floor plane selected inside the pocket   Fig.4-18 Floor plane selected as the bottom of the part

4.3.2 Planar milling islands and pockets

NX regards areas enclosed by part boundaries with material retained inside as islands. This means that some areas not conventionally thought of as islands are regarded as islands by NX. It is important to keep this definition in mind when using depth of cut parameters options such as floor & island tops, levels at island tops, and top off islands to understand the behavior.

Fig.4-19 shows the outside periphery defined by a part boundary, and the material is placed outside. In the other boundaries, the material is inside. These are island boundaries.

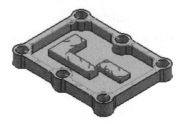

Fig.4-19 Outside periphery defined by a part boundary

4.4 Planar milling operation parameters

4.4.1 Cut patterns

Cut patterns in planar operations determine the tool path pattern used to machine cut regions. Eight cut patterns are available in planar milling, as shown in Fig.4-20.

Fig.4-20 Cut patters in planar milling

1. Zig

The zig cut pattern always cuts in one direction (see Fig.4-21). The tool retracts at the end of each cut, then moves to the start position for the next cutting pass. Climb (or Conventional) cutting is maintained.

In the zig cut pattern, tool passes have the following characteristics.

· They always cut in one direction.

· They do not perform contour cutting between consecutive passes unless they are required to do so by the specified engage method.

· They engage from open air unless doing so would cause the tool to move across a large portion of the cut region that was previously machined. In that case, the pattern re-engages at the start of the previous zig pass to minimize the amount of tool travel along the cut portion of the region.

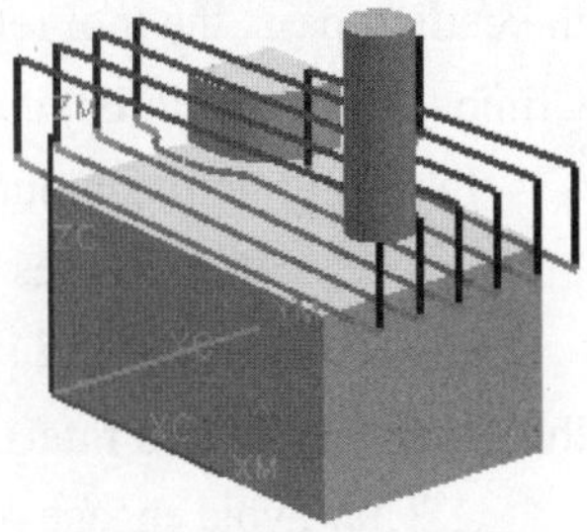

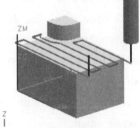

Fig.4-21 Zig cut pattern

· They follow cut region contours to maintain a continuous cutting motion as long as the passes. So do not intersect. They only deviate from a straight line path by less than the stepover value.

· They are shortened if there is an obstruction.

To prevent leaving too much material along the part walls at the next cut level, the wall cleanup option is recommended.

2. Zig-zag

The zig-zag cut pattern (see Fig.4-22) machines in a series of parallel straight line passes that cut in opposite directions while stepping over in one direction. This cut pattern allows the tool to remain continually engaged during stepovers.

In the zig-zag cut pattern, tool passes have the following characteristics:

Fig.4-22 Zig-zag cut pattern

· They start as close as possible to the start point of the peripheral boundary unless you specify region start points.

· They follow cut region contours to maintain a continuous cutting motion as long as the passes, so do not intersect. They only deviate from a straight line path by less than the stepover value.

· They have stepover moves that always follow cut region contours.

· They are shortened if there is an obstruction.

The cut direction settings (climb cut or conventional cut) are ignored since the cut direction changes from one pass to the next. To prevent leaving too much material along the part walls at the next cut level, the Wall Cleanup option is recommended.

3. Zig with contour

The zig with contour cut pattern (see Fig.4-23) machines with cuts going in one direction. Contouring moves along the boundary added before and after the linear passes. At the end of

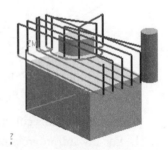

Fig.4-23 Zig with contour cut pattern

the cutting pass, the tool retracts and re-engages at the start of the contouring move for the next cut. Climb (or Conventional) cutting is maintained.

In the zig with contour cut pattern, tool passes have the following characteristics.

· They cut in a series of loops that maintain a climb or conventional cut direction.

· They have cutting stepover moves that follow the cut region contours. Stepover moves use the specified cut feed rate and ignore the stepover feed rate from the Feeds and speeds dialog box.

· They avoid engage moves into the corner of a cavity or along a wall to produce a cleaner cut.

To prevent leaving too much material along the part walls at the next cut level, the wall cleanup option is recommended.

4. Follow periphery

The follow periphery cut pattern (see Fig.4-24) cuts along offsets from the outermost edge that is defined by part or blank geometry. Internal islands and cavities require island cleanup or a clean up profile pass. Climb (or Conventional) cutting is maintained.

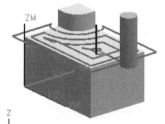

In the follow periphery cut pattern, tool passes have the following characteristics.

(1) They are closed shapes offset from the peripheral loop of the cut region. In narrow regions, the software adds a finish pass, and adjusts the follow periphery passes to leave material to be cut by the finish pass.

Fig.4-24 Follow periphery cut pattern

(2) They merge with island shapes when it is no longer possible to continue cutting with the periphery shape.

(3) They follow the cut region contours to maintain a continuous cutting motion as long as the passes do not intersect. If one area requires more pocketing passes than another area, the software adds a transfer move between the subregions to complete the larger area as shown in Fig.4-25.

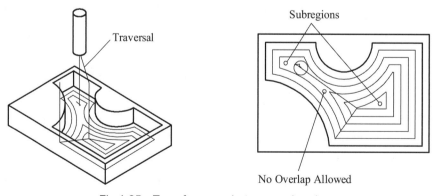

Fig.4-25 Transfer move between subregions

(4) They follow this cut order:

① For inward progression, cuts all the open passes first, and then cuts all the closed inner passes.

② For outward progression, cuts all the closed inner passes first, and then cuts all the open passes.

To cut open passes, the tool engages to an open pass, cuts, retracts, and then transfers to another open pass.

(5) They can generate diagonal tool movements, called tusks, to remove material from the corners.

Tusks (see Fig.4-26) are necessary when the stepover is large in relation to the tool diameter and there is no overlap between passes at the corners.

(6) They can generate additional cleanup moves in areas between consecutive passes.

Additional cleanup moves are sometimes necessary for large stepovers. A stepover is considered large if it is greater than 50% of the tool diameter, but less than 100% of the tool diameter.

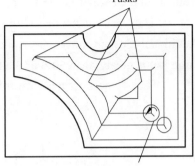

Additional Movement along tusk cleans out material between corners

Fig.4-26 Tusks

5. Follow part

The follow part cut pattern (see Fig.4-27) cuts along concentric offsets from all specified part geometry. The outermost edge and all interior islands and cavities are used to compute the tool path. This eliminates the need for an island cleanup pass. Climb (or Conventional) cutting is maintained.

The follow part cut pattern is recommended for cavity regions with islands.

In the follow part cut pattern, tool passes have the following characteristics.

· They are offset equally from all part geometry, including the peripheral loop, islands, and cavities. Intersecting offsets do not cross, but trim to one another as shown in Fig.4-28.

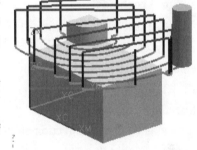

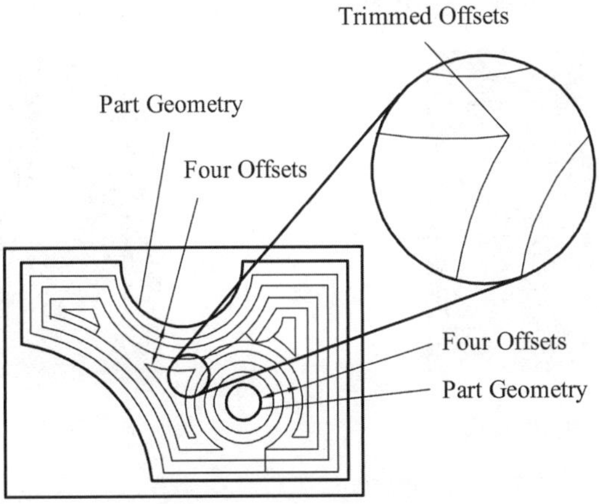

Fig.4-27 Follow part cut pattern

Fig.4-28 Trimmed offsets

• They are offset from blank geometry only when there is no defined part geometry to offset from, as in the case of facing regions.

• They can generate additional cleanup moves in areas between consecutive passes. Additional cleanup moves are sometimes necessary for large stepovers. A stepover is considered large if it is greater than 50% of the tool diameter, but less than 100% of the tool diameter.

• They always progress towards the part geometry.

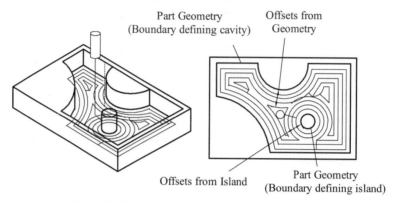

Fig.4-29   Tool passes with follow part cut pattern

6. Trochoidal

Use the Trochoidal cut pattern to limit excess stepover to prevent tool breakage when the tool is fully embedded into a cut, or the tool should avoid embedding. Most cut patterns generate embedded regions between islands and parts during the engage as well as in narrow areas.

There are significant differences between the outward and inward cut directions for the trochoidal cut pattern.

• The outward direction starts away from part walls and progresses towards the part walls as shown in Fig.4-30. This is the preferred pattern, and efficiently combines a circular loop and smooth follow movement.

• The inward direction cuts along the part in loops, and then cuts the inward passes in a smooth follow periphery pattern as shown in Fig.4-31.

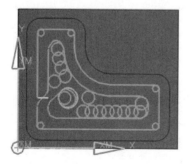 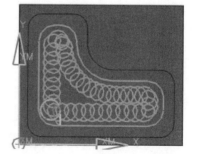

Fig.4-30   Outward trochoidal   Fig.4-31   Inward trochoidal

7. Profile

The Profile cut pattern (see Fig.4-32) machines along part walls with the side of the tool to create a finishing pass. The tool follows the boundary direction.

In the Profile cut pattern (see Fig.4-33), tool passes have the following characteristics.

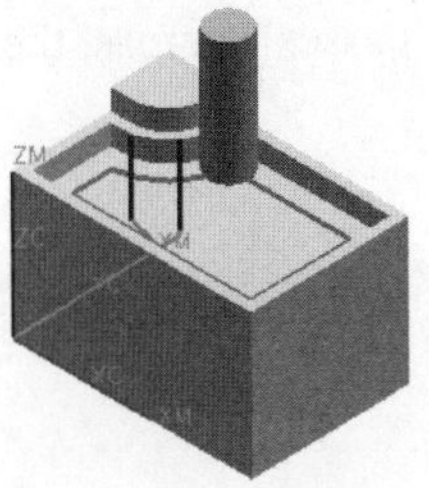

· They can machine both open and closed regions. Closed shapes are built and traversed in the same way as the follow part cut pattern.

· They can use a custom tool, such as a chamfering tool.

· They can be limited to small areas by defining blank geometry only in the area of the part you want to finish, as shown in Fig.4-34.

Fig.4-32 Profile cut pattern

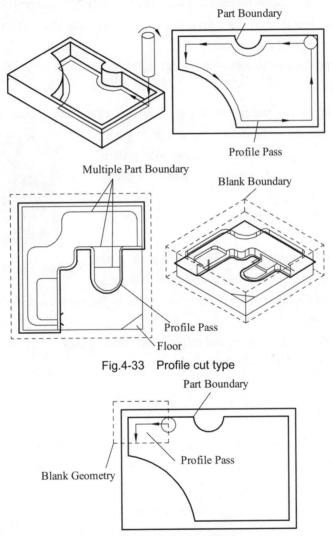

Fig.4-33 Profile cut type

Fig.4-34 Finishing a small area

· They require part boundaries that do not cross so that the software can determine the material side.

· They can cut more than one open region at a time. If the open regions are close enough for tool passes to cross, the software adjusts the tool path as shown in Fig.4-35.

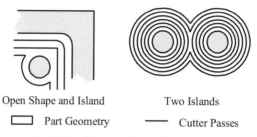

Open Shape and Island   Two Islands
☐ Part Geometry   ── Cutter Passes

Fig.4-35 Overlapping cutter passes

· They can generate additional cleanup moves in areas between consecutive passes. Additional cleanup moves are sometimes necessary for large stepovers (greater than 50% of the tool diameter, but less than 100% of the tool diameter).

· They do not produce cutting motions in cut regions that consist entire blank geometry.

8. Standard drive

The standard drive cut pattern (see Fig.4-36) creates profiling cuts along the specified boundaries without automatic boundary trimming or gouge checking. You can specify whether or not the tool path is allowed to cross itself. This cut pattern is available in planar milling only.

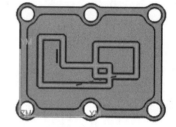

In the standard drive cut pattern, tool passes have the following characteristics.

· They treat each shape as a separate region. The software traces each shape independently, and cleans up each shape independently of the other shapes. Boolean operations are not performed between shapes.

Fig.4-36 Standard drive cut pattern

· They can intersect.

· They do not check for gouging, and result in overlapping cats.

· They ignore all check and trim boundaries.

4.4.2 Cut levels

Cut levels are used to specify cut ranges and the depth of the cut within each range. The cut levels of planar milling are defined in path setting groups, as shown in Fig.4-37.

Fig.4-37 Cut levels dialog box

The depth of cut parameters can be defined with five types as following.

1. User Defined

Lets you enter parameters for depth of cut using.

· Maximum — Maximum defines the largest allowable cut depth for each cut level occurring after the initial level and before the final level.

· Minimum — Minimum defines the smallest allowable cut depth for each cut level occurring after the initial level and before the final level.

· Initial —Initial defines the cut depth for the first cut level. This value is measured from the blank boundary plane and is independent of the maximum and minimum values.

· Final — Initial defines the cut depth for the last cut level. This value is measured from the floor plane.

2. Floor only

Lets you create a tool path only at the floor level.

3. Floor & island tops

Lets you create a tool path at the final depth followed by clean up paths at the top of each island.

4. Levels at island tops

Lets you create a planar cut at the top of each island, this path will cut completely at each level before moving to the next deeper level. This option works from the top boundary to the final floor plane.

5. Fixed depth

Lets you select the maximum depth of cut.

You can add additional stock at each cut level in increment side stock box, which allows clearance for tools with short flute lengths but does not remove all the stock from the boundaries. The top off islands checkbox lets you generate a separate path on the top of each island that the processor can not initially clean with one of the cut levels.

4.5 Planar milling cutting parameters

4.5.1 Wall cleanup

Wall cleanup is available for the zig, zig-zag, and follow periphery cut patterns in planar milling. It inserts a final profile pass at each cut level to remove ridges that remain along the part walls. Wall cleanup passes are different from profile passes.

· A wall cleanup pass is used for roughing while a profile pass is a finishing move.

· A wall cleanup pass uses the part stock while a profile pass uses the finish stock to offset the tool path.

• A wall cleanup pass inserts a final profile pass at each cut level while a profile pass cuts only at the floor level.

There are four options in the wall cleanup list.

1. Automatic

Available for the follow periphery cut pattern as shown in Fig.4-38. Uses profile passes to remove all material without re-cutting the material.

2. None

Do not always remove all the material, but can create a shorter tool path with fewer engages as show Fig.4-39.

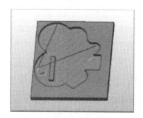
Fig.4-38 Automatic

Fig.4-39 None

3. At start/ end

Makes an extra profile pass along the part walls to remove uncut material and re-cut some of the outer follow periphery pass as shown in Fig.4-40.

4. Cut walls only

Limits the cut path to walls only as shown in Fig.4-41.

Fig.4-40 At start/ End

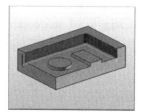
Fig.4-41 Cut walls only

4.5.2 Finish passes

Finish passes controls the last cutting pass (or passes) the tool makes after completing the main cutting passes. It's available for the following cut patterns in planar milling.

• Follow part.
• Follow periphery.
• Trochoidal.
• Zig.
• Zig-zag.

· Zig with contour.

You can add one or more finish passes (see Fig.4-42) to the operation and it requests multiple finish passes with centerline cutter compensation.

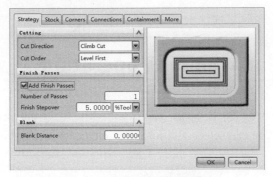

Fig.4-42　Add finish passes

· Number of passes: Specifies the number of finish passes to add.

· Finish stepover: Specifies the stepover value to apply only to the finish passes as shown in Fig.4-43.

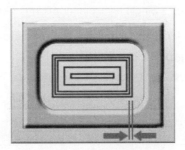

Fig.4-43　Finish stepover

4.5.3　Final floor stock

Final floor stock (see Fig.4-44) is available for face milling and planar milling operations. It sets a value for the amount of material to remain uncut. The floor stock is measured from the face plane and offset along the tool axis. In planar milling, material remains uncut on the pocket floors and island tops after the tool path is complete. In face milling, material remains uncut on face geometry after the tool path is complete.

Fig.4-44　Final floor stock

103

You can enter a negative value when you want to make a minor adjustment to a modeled dimension in order to meet a functional requirement. For example, a negative value is useful when you want to shorten the height of a face or move a face into a part.

4.6 Create planar milling operation

This example shows the basic steps you would use to create a single level planar mill operation (see Fig.4-45). You can select additional parameters to create the desired tool path.

Fig.4-45 Single level planar milling operation

(1) On the insert toolbar, click Create operation or choose insert→operation.
(2) In the operation subtype group, click Planar mill .
(3) In the location group, select the following (see Tab.4-4).

Tab.4-4 Operation options

Program	Program
Tool	EM-.75-.03
Geometry	Workpiece
Method	Mill rough

(4) Click OK. The Planar mill dialog box is displayed.
(5) In the geometry group, click Specify part boundaries .
(6) In the Boundary geometry dialog box, select the Ignore holes check box.
(7) Select the top face of the part to be machined (see Fig.4-46).

Fig.4-46 Select the top face

(8) Click OK. The Planar mill dialog box is displayed.

(9) The boundary is created (see Fig.4-47) and displayed, and the holes are ignored as specified.

(10) Click Specify floor .

(11) Select the floor geometry (see Fig.4-48).

Fig.4-47　The boundary is created　　　　Fig.4-48　Select the floor geometry

(12) This is the final depth of the path.

(13) Click OK to return to the Planar mill dialog box.

(14) In the path settings group, from the cut pattern list, select Profile.

(15) Click Generate .

(16) The tool path is generated and displayed.

4.7　Planar milling example

This example shows the basic steps you would use to create a multi level planar mill operation, and select additional parameters to create the desired tool path. The part to be machined is shown in Fig.4-49.

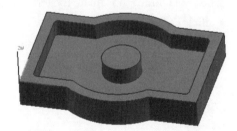

Fig.4-49　The part to be machined

4.7.1　Create rough planar milling operation

1. Setting up the machining environment

(1) Choose start→manufacturing.

105

(2) In the Machining environment dialog box, CAM session configuration group, select CAM general.

(3) In the CAM setup to create group, select mill planar.

(4) Click OK.

2. Create machine tool

(1) On the insert toolbar, click Create tool, or choose insert→tool.

(2) In the Create tool dialog box, from the type list, select mill planar. In the tool subtype group, click mill. Set machine tool name as Mill_D10.

(3) The software displays the Five parameter mill tool dialog box.

(4) In the Five parameter mill tool dialog box, set the tool parameters as shown in Fig.4-50.

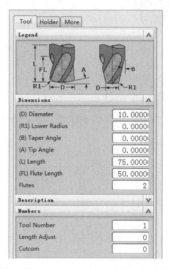

Fig.4-50　Five parameter mill tool dialog Box

3. Setting geometry

(1) In the Geometry order view, double click MCS mill, the Mill orient dialog box appears.

(2) In the Mill orient dialog box, Machine Coordinate System group, click CSYS dialog.

(3) In the CSYS dialog box, click Dynamic.

(4) In the graphics window, use the handles on the MCS to move it to the end point of the top surface, as shown in Fig.4-51.

 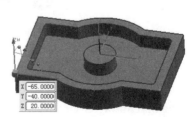

Fig.4-51　Set MCS

(5) Click OK to return to the MCS dialog box.

(6) Click OK.

(7) In the geometry order view, double click Workpiece, the Mill geom dialog box appears.

(8) In the Mill geom dialog box, click Specify part . The Part geometry dialog box opens.

(9) In the graphics window, select the entire model, as shown in Fig.4-52.

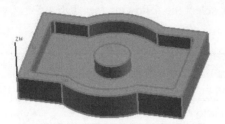

Fig.4-52 Select part geometry

(10) Click OK to close the Part geometry dialog box.

(11) In the Mill geom dialog box, click Specify blank . The Blank geometry dialog box opens.

(12) In the graphics window, select the blank, as shown in Fig.4-53.

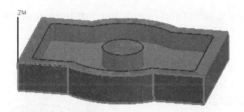

Fig.4-53 Specify Blank Geometry

(13) Click OK to close the Blank geometry dialog box.

(14) Click OK. The workpiece is specified.

4. Setting machining method

(1) In the machining method order view, double click Mill rough. The Mill method dialog box opens.

(2) In the Mill method dialog box, set parameters as shown in Fig.4-54.

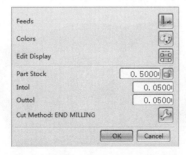

Fig.4-54 Set mill method

(3) Click Ok.

5. Create operation

(1) On the insert toolbar, click Create operation or choose insert→operation.

(2) In the operation subtype group, click Planar mill .

(3) In the location group, select the following (see Tab.4-5).

Tab.4-5 Operation options

Program	program
Tool	Mill_D10
Geometry	Workpiece
Method	Mill rough

(4) Click OK. The Planar mill dialog box is displayed.

(5) In the geometry group, click Specify part boundaries as shown in Fig.4-55.

(6) In the Boundary geometry dialog box, clear the Ignore islands check box in the face selection group.

(7) In the graphic windows, select the face of the part to be machined.

Fig.4-55 Specify part boundary

(8) Click OK. The Planar mill dialog box is displayed. The boundary is created and displayed, and the island is not ignored as specified as shown in fig.4-56.

Fig.4-56 Part boundary

(9) In the geometry group, click Specify blank boundaries as shown in Fig.4-57.

(10) In the Boundary geometry dialog box, clear the Ignore Islands check box and select Ignore hole check box in the face selection group.

(11) In the graphic windows, select the top face of the part.

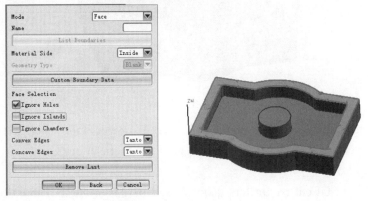

Fig.4-57　Specify blank boundary

(12) Click OK. The Planar mill dialog box is displayed. The boundary is created and displayed as shown in Fig.4-58.

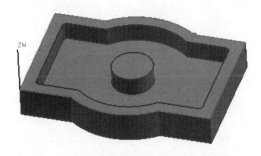

Fig.4-58　Blank boundary

(13) In the geometry group, click Specify floor. The Plane constructor dialog box appears as shown in Fig.4-59.

(14) In the graphic windows, select the bottom face of the internal cavity of the part.

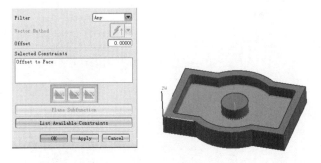

Fig.4-59　Specify floor

109

(15) Click OK. The Planar mill dialog box is displayed. The Floor is created and displayed as shown in Fig.4-60.

Fig.4-60 Floor

(16) In the path settings group, from the cut pattern list, select follow periphery. Select % tool flat from the stepover list, and type 50 in the Percent of flat diameter box .

(17) Click Cut levels.

(18) In the Depth Of cut parameters dialog box, type 2.0 in the maximum box as shown in Fig.4-61.

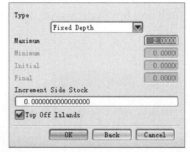

Fig.4-61 Set depth of cut parameters

(19) Click OK to return to the Planar mill dialog box.

(20) In the path settings group, click Cutting parameters (see Fig.4-62), opens the Cutting parameters dialog box.

(21) In the Strategy tab, select Island cleanup check box.

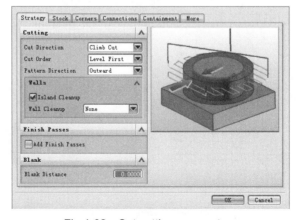

Fig.4-62 Set cutting parameters

110

(22) Click OK to return to the Planar mill dialog box.

(23) Click Generate .

(24) The tool path is generated and displayed as shown in Fig.4-63.

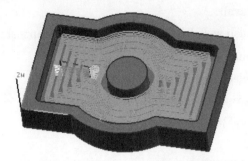

Fig.4-63 The tool path of a rough planar mill operation

4.7.2 Create finish planar milling operation

1. Create machine Tool

(1) On the Insert toolbar, click Create tool , or choose insert→tool.

(2) In the Create tool dialog box, from the type list, select mill planar. In the tool subtype group, click Mill . Set machine tool name as Mill_D5.

(3) The software displays the Five parameter mill tool dialog box.

(4) In the Five parameter mill tool dialog box, set the tool parameters as shown in Fig.4-64.

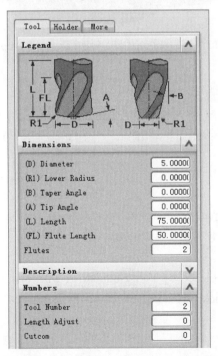

Fig.4-64 Five parameter mill tool dialog box

2. Copy operation

(1) In the program order view, copy the planar mill operation which is previously created, and paste to the program node.

(2) Rename it as finish wall.

(3) Double click the finish wall, The Planar mill dialog box appears.

3. Set operation parameters

(1) In the tool group, select Mill D5 from Tool list.

(2) In the path setting group, select mill finish from method list.

(3) Select profile from cut pattern list.

(4) Click Cut levels.

(5) In the Depth of cut parameters dialog box, type 1.0 in the maximum box.

(6) Click OK to return to the Planar mill dialog box.

(7) Click Generate.

(8) The tool path is generated and displayed as shown in Fig.4-65.

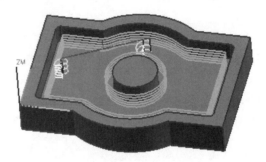

Fig.4-65 The tool path of a finish planar mill operation

 Notes

[1] Permanent boundaries continually display on the screen like any other permanent geometric entity (line, arc, etc.) and can be used in any machining module requiring boundaries.

永久边界在屏幕上持续显示，如同其他永久的几何实体（直线，圆弧等），并且可以在需要边界的任何加工模型中使用。

[2] Trim boundaries are used to limit the cut regions at each cut level. Combine trim boundaries with the specified part geometry to discard the cut region outside of the trim boundary.

修剪边界用于限制在所有切削层中的切削区域。修剪边界与特定的部件几何体结合以摒弃修剪边界之外的切削区域。

[3] NX regards areas enclosed by Part Boundaries with material retained inside as islands. This means that some areas those are not conventionally thought of islands are regarded as islands by NX.

NX 将由保留在内部的材料的部件边界围绕的区域视为岛屿。这就意味着 NX 会把一些并不传统的区域也看作为岛屿。

New Words in Chapter 4

permanent boundary			永久边界
temporary boundary			临时边界
chamfer	['tʃæmfə]	n.	斜面，凹槽
underneath	[ʌndə'ni:θ]	prep.	在…的下面
intersect	[ɪntə'sekt]	vi.	相交，交叉
peripheral	[pə'rɪfərəl]	adj.	外围的，外部的
obstruction	[əb'strʌkʃ(ə)n]	n.	障碍，阻碍
subregion	['sʌbˌrɪdʒən]	n.	子区域
tusk	[tʌsk]	n.	尖头，尖形物
concentric	[kən'sentrɪk]	adj.	同轴的，同中心的
breakage	['breɪkɪdʒ]	n.	破坏，破损

Chapter 5
Cavity Milling

Objectives:
- ✓ To understand the cavity milling.
- ✓ To understand the IPW.
- ✓ To be able to create a cavity milling.

5.1 Cavity milling overview

5.1.1 Cavity milling operation introduction

Use cavity mill operations (see Fig.5-1) to remove large volumes of material. Cavity mill is ideal for rough-cutting parts, such as dies, castings, and forgings. Cavity mill removes material in planar levels that are perpendicular to a fixed tool axis. Cutting is completed at a constant z-level before moving onto the next Z-level. Part geometry can be planar or contoured.

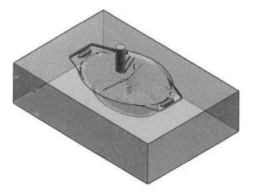

Fig.5-1 Cavity milling

In cavity milling, you must first select or define part and blank geometry. The software then:
- Sets the top and bottom of the blank geometry at the highest and lowest level of cutting.
- Creates one or more planes that are perpendicular to the tool axis, at the defined cut level.
- Creates intersection curves or traces between the cut level planes and the geometry.

- Creates a cut pattern at each cut level.
- Combines engage and retract moves with the different cut levels.

You can have a single cavity milling operation, or a series of cavity milling operations that take advantage of the In Process Workpiece (IPW). When you use an IPW, the software:
- Captures what has been cut or not been cut.
- Provides advanced functionality for tool holder collision checking.
- Allows you to view the material removed by the current operation.

5.1.2 Cavity milling operation subtypes

The cavity milling operation subtypes are shown in Tab.5-1.

Tab.5-1 Cavity milling operation subtypes

Icon	Subtype	Description
	Cavity mill	The main cavity milling operation subtype
	Corner rough	Customized to cut the material remaining in corners that a previous tool could not reach, due to its diameter and corner radius
	Rest milling	Customized to cut the material remaining from the IPW that a previous tool could not reach, due to its diameter and corner radius

5.2 In Process Workpiece (IPW)

5.2.1 IPW overview

The In Process Workpiece (IPW) (see Fig.5-2) is a geometric shape that is produced by the manufacturing application to represent the machined workpiece at each stage of machining. According to the order of the operations in the program order view, each operation tool path progressively reduces the IPW to mimic the material removed from the actual workpiece on the machine tool. This provides exceptional benefits to the user and to the tool path processors. For tool path generation, some of the operations can use the IPW shape from the previous operation as input, which reduces user interaction and improves tool path cutting efficiency. For verification and simulation, the IPW is progressively modified by all tool paths throughout turning, milling, and drilling operations (with very few exceptions).

Manufacturing creates different types of IPWs

Fig.5-2 In Process Workpiece

depending on the type of operation that you use.

(1) Milling/Drilling — use a non-rotating IPW. This type of IPW is a three dimensional representation of the state of material removal in an operation at the point machining is stopped.

(2) Turning — uses a spinning (rotating) IPW. This is equivalent to a static IPW that is spun around the axis of rotation of the lathe spindle.

(3) Mill turn — when generating a mill-turn program, the software maintains the IPW for each milling and turning operation, and the conversion between the non-rotating and rotating states of the IPW is done automatically by the software.

In the case where a turning operation follows a milling operation, the software spins the static IPW of the milling operation around the axis of rotation of the turning operation's spindle coordinate system. It then creates a 2D cross-sectional silhouette of the spinning IPW and uses that spinning silhouette as the blank input to the turning operation (see Fig.5-3).

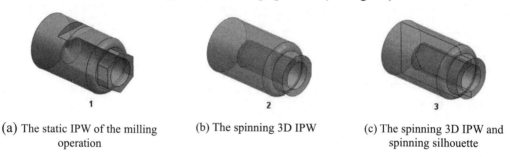

(a) The static IPW of the milling operation   (b) The spinning 3D IPW   (c) The spinning 3D IPW and spinning silhouette

Fig.5-3 Different types of In Process Workpiece

In the case where a milling operation follows a turning operation, the software uses the dynamic IPW of the turning operation as the spun outline of a body of revolution. This represents the current state of the machined part in a spinning situation. The steps taken here depend on whether the turning operation is the first operation of the mill-turn program or not.

· If it is the first operation, the blank for the milling operation is the intersection of the dynamic IPW with the original blank geometry of the turning operation.

· If it is not the first operation, the blank for the milling operation is the intersection of the dynamic IPW with the static IPW of the milling operation that preceded the turning operation.

For milling operations, you must define the initial workpiece and specify cutting parameters options to create the IPW.

The benefits of using IPW are:

· The processor machines the part based on the current state of the real workpiece. This avoids cutting air in the regions which have already been cut.

· Successive operations can use smaller radius tools to machine the areas that were not cut by the larger tools on prior operations.

· Successive operations can use tools of similar shape but increasing length to cut the most material with the shortest tool. The follow up operations with the longer tool lengths only machine material that was unreachable of the other operations.

· In cavity milling, you can display both the input and output 3D IPW for an operation. From the Operation Navigator, you can display the output 3D IPW for an operation.

5.2.2 How processors use the IPW

IPW allows the software to recognize the material remaining from prior operations. The IPW is used in the following ways:

· To visualize the material remaining from prior operations. This is the input state of the IPW.

· To define the blank.

Operations that use a blank to determine the volume of material to cut (cavity milling, plunge milling), can also use the input IPW from prior operations to define the blank. These are commonly called rest mill operations because they cut only the "rest" of the material.

· To check for tool collisions.

Some operations that don't support a blank (like hole making) can use the input IPW for global collision avoidance against the tool.

· To avoid tool holder collision regions.

Operations that support IPW as blank (like cavity milling), will use the dynamic IPW that updates after each region is cut to ensure that tool holder collision regions are not cut. This allows for maximum depth of cutting while avoiding tool holder collision regions.

· To output the material remaining from the current operation.

There are several IPW containment options as shown in Fig.5-4.

· Use Level-based option creates 2D representations of uncut regions for cavity milling and Z-level milling operations.

· Use 3D option creates a facet body to represent the IPW.

· Use 2D IPW option creates 2D representations of uncut regions for planar milling operations.

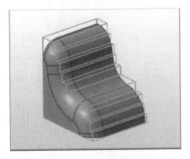 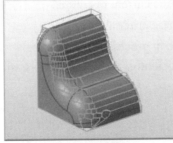 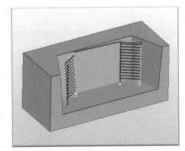

(a) Use Level-Based   (b) Use 3D   (c) Use 2D IPW

Fig.5-4 IPW Options

Each processor uses the IPW differently, as shown in Tab.5-2.

117

Tab.5-2 How processors use the IPW

Operation	How it uses the IPW	Containment options
Planar milling	Inputs IPW as blank.	Use 2D IPW in previous and current operations
Z-level: Profiling, Surface contouring: Area Milling, and Flowcut	Only machines regions that were uncut because of prior tool holder collisions	· Use tool holder in previous operation · Use 2D workpiece in current operation
Cavity milling and rest milling	Defines the input blank	IPW · Use 3D option · Use level-based
	Tool holder avoidance, dynamic IPW	Use tool holder in previous and current operations
Surface contouring: variable axis contour profile	Collision checking	IPW · Use 3D option
Hole making	Collision checking	IPW · Use 3D option · Use level-based option

1. Use level based option

The use level-based option is available for cavity milling, plunge milling, and rest milling operations to create representations of uncut regions. Level-based IPW uses 2D cut regions from previous operations that are referenced to identify the remaining stock. Fig.5-5 shows the resulting tool path from level-based IPW.

Use level-based IPW is for corners and stair-steps on walls. This option does not consider blank material left by adjacent passes within a cut level pattern. Scallops within the cut-regions are ignored and considered fully removed.

The reference operations used to create the level-based IPW include cavity milling, Z-level milling, face milling, or planar milling operations. These operations:

· Are in the same geometry group as the level based operation. The geometry group must have blank geometry defined.

· Use the same tool axis.

· Use a tool at least as big as the current operation.

Operations that do not meet these criteria are ignored.

Benefits of using a level-based in process workpiece are:

· Use level-based IPW efficiently cuts the corners and stair-steps left from previous operations.

· Tool path processing time is noticeably shorter than Use 3D IPW for simple parts, and dramatically shorter for larger complex parts.

· You can use a big cutter with big depths of cut in one operation, and then use the same tool and a much smaller depth of cut in the following operation to clean out stair-steps.

· The tool path is much cleaner than using the Use 3D IPW option.

· You can further automate by combining multiple roughing operations to rough and rest mill a given cavity.

· You can combine this option with the use tool holder option so that shorter length tools (that are more rigid) can be used to cut to greater depths within a cavity. The longer length tool in the following operation only needs to cut along the walls where the shorter tool could not reach because of holder violation.

Fig.5-5 Resulting tool path from level-based IPW

2. Use 3D

The use 3D option creates a 3D facet body to represent the remaining material. The facet body can have many small specks of material and require large amounts of memory to create. The use level-based option creates a 2D IPW. This option is preferred for its faster performance and cleaner tool paths.

In cavity milling and plunge milling operations, select the use 3D option when:

· You are roughing with reference operations other than cavity milling, face milling, Z-level milling or planar milling.

· Any of the reference operations have a different tool axis than that of the current operation.

· Any of the reference operations have a smaller tool size than that of the current operation.

· You want to use the IPW from the current setup in another setup.

3. Use 2D IPW

Use the use 2D IPW option in planar milling to avoid tool motion where material has been removed. Subsequent operations machine only the material that is left on the part. The software tracks the remaining material from one operation to the next.

When you use this option, the software:

· Finds uncut material by evaluating previous planar milling operations that share the same part geometry.

· Only recognizes valid planar milling operations including planar profile, profile 3D, solid profile 3D, and planar mill. If other operations are found, the software ignores them and uses the 2D IPW from the last planar milling operation.

· Supports the use of ball end mills.

In Fig.5-6, the first tool path roughs out most of the material, while the second tool path removes only the remainder of the material recognized in the IPW.

(a) The first tool path   (b) The second tool path

Fig.5-6   Use 2D IPW

5.2.3 Visualizing the IPW

Using Visualize, you can see the progress of the IPW from the initial blank to the finished workpiece. The initial blank defined in the workpiece geometry group is shown in Fig.5-7. IPW after the first operation is shown as Fig.5-8.

Fig.5-7　The initial blank　　　　　　　　Fig.5-8　IPW after the first operation

Second operation cutting with IPW option is shown in Fig.5-9, and the finished part is shown in Fig.5-10.

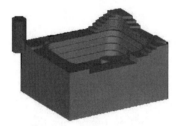

Fig.5-9　The second operation cutting with IPW　　　　Fig.5-10　Finished part

5.2.4　Save IPW

You can save each IPW as the default by following:
(1) Choose file→utilities→customer defaults.
(2) In the Customer defaults dialog box, select manufacturing→geometry.
(3) Click the IPW tab.
(4) Select the save in separate part file option.
(5) Select the type of IPW you want to save in a separate part file.
① To save each 3D IPW, select the IPW model check box.
② To save each level based IPW, select the level-based IPW check box.
(6) Click OK to close the dialog box.
(7) Click OK to dismiss the message.
This setting takes effect when you restart NX the next time .

If you want to save each IPW in a separate part file for the current session, you can do as follows.

(1) Choose preferences→ manufacturing → IPW model.

(2) In the Manufacturing preferences dialog box, click the Configuration tab.

(3) In the configuration tab, select the save in separate part files option.

① To save each 3D IPW, select the IPW model check box.

② To save each level-based IPW, select the level-based IPW check box.

(4) (Optional) Clear the Use directory of original part check box and specify which folder to save the IPW in.

(5) Click OK.

5.3 Setting cavity milling geometry

5.3.1 Valid geometry for cavity milling

You can specify the following types of geometry for cavity milling operations, as shown in Tab.5-3.

Tab.5-3 Valid geometry for cavity milling

Icon	Geometry	Description
	Part	Specify the geometry to machine for roughing and finishing operations
	Blank	Define the raw stock or the material to be removed during machining
	Check	Specify geometry that you want the tool to avoid. An example of check geometry would be clamps that hold the part
	Cut Area	Specify the areas of your part to machine
	Trim boundaries	Limit the cut regions at each cut level

5.3.2 Blank geometry

Use blank geometry in cavity milling (see Fig.5-11) to define the raw stock or the material to be removed during machining. Cavity milling can save the remaining stock to be used in subsequent operations. This is referred as the IPW.

You can also use blank geometry to isolate areas of the part that are to be machined as shown in Fig.5-12.

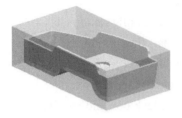 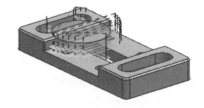

Fig.5-11　Blank geometry for cavity milling　　Fig.5-12　Using blank geometry to isolate areas of the part

It is not always necessary to specify blank geometry in cavity milling. The Part geometry alone may encloses and defines the entire cut volume as shown in Fig.5-13.

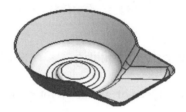

Fig.5-13　The part geometry defines the entire cut volume

You can define blank geometry in the following ways:

(1) Select geometry using faces, curves, facet bodies or solid bodies. When you select curves, the system extends the curve to the lowest level along the taper angle.

(2) Define a blank distance (Cutting parameters dialog box→Strategy tab). Blank distance creates an internally defined offset of the part geometry (see Fig.5-15).

· Specify either auto block or part offset in the geometry parent.

· The finished part is shown as Fig.5-14.

Fig.5-14　Finished part　　　　　　　　　Fig.5-15　Blank Distance offset

5.3.3　IPW as blank geometry

In cavity milling operations, the IPW options control what is input as the initial blank from prior operations within the same geometry group. They also control the output, which represents the state of the workpiece after completing the operation.

1. Requirements to use an IPW as blank

Before you can use an IPW as blank geometry, the operation must satisfy certain requirements.

• The operation must be within a geometry group (Workpiece or Mill Geom) and have defined blank geometry.

• All the operations in the same geometry group and program view hierarchy have a valid tool path.

If an operation does not satisfy the requirements, the software displays error messages to help you correct the problem when you generate the tool path.

2. Improving performance when using the IPW as blank

Using and displaying a 3D IPW requires a large amount of memory to create the representation. For complex parts, a level-based IPW may also require a large amount of memory. To reduce the required memory and reuse 3D geometry, save the IPW in separate (component) part files for each operation.

• The software adds a reference set to the assembly and names it <operation name>.

• The software creates component parts and names the component parts <part name>_ipw_<operation name>. prt. The search path used for finding the component part is what is supplied in the preference option instead of general assembly search option.

The component parts store:

• The 2D cut regions (2D IPW) for an operation with the use level-based option . These cut regions are not viewable in the component part file.

• The resultant facet body (3D IPW) for an operation with the Use 3D option. The geometry can be reused as long as the operation is up to date. The 3D IPW is viewable in the component part file.

3. IPW behavior for operations that use the IPW as blank

When you select the use 3D option or use level-based option in cavity milling or plunge milling dialog box, and generate the operation, the software searches the geometry view hierarchy in the Operation Navigator to ensure that it meets the requirements of a valid IPW.

When an operation with a valid IPW is not found in the program order view hierarchy, but all the preceding operations in the sequence have a valid tool path, the software:

(1) Applies the tool paths for all previous operations to the blank defined in the parent geometry group to create the reference IPW geometry (previous IPW).

(2) Inputs the reference IPW instead of the defined blank geometry.

(3) Applies the tool path for the current operation to the reference IPW.

(4) Stores the resulting In Process Workpiece with the current operation, or in component files if the save in separate part files option is selected.

When an operation with a valid IPW is found in the sequence, the software:

(1) Ignores any inherited blank geometry. The software does not allow the use of blank geometry and the IPW option at the same time in any operation.

(2) Applies the tool paths from all the subsequent operations up to the current operation, to the valid IPW to create the new reference IPW (Previous IPW).

(3) Applies the tool path for the current operation to the reference IPW.

(4) Stores the resulting In Process Workpiece with the current operation, or in component files if the save in separate part files option is selected.

When the IPW option is changed for an operation, the tool path status of the operation is changed to regenerate. When an operation has a status of regenerate, the subsequent operations in the geometry view and program view hierarchy are also marked as regenerate if they:

· Refer to the same initial workpiece.

· Have the IPW use 3D option or use level-based option selected, or have an IPW object generated.

Operations within the unused items group in the program order view of the Operation Navigator are ignored by Visualize and Cavity milling. These operations:

· Do not contribute to the IPW for any operations in the same geometry view hierarchy.

· Use the originally defined blank geometry regardless of the IPW option setting.

5.3.4 Cut area

For the best results in cavity milling, Cut area geometry should contain faces that clearly define a volume of material to remove.

If the cut area contains only vertical faces, the system still attempts to define a volume to remove, but it may not be what you expect. By selecting more faces, adjusting the extension distance, or changing the blank, you should be able to get the desired tool path for a local area.

If the cut area contains only horizontal faces, then no depth of the volume is defined, and the cut levels may be extended up to the top of the blank. To limit the cut depths to the local area, either select some non-horizontal faces to represent the depth, or adjust the cut levels manually.

5.4 Cavity milling cutting parameters

5.4.1 Cut levels overview

Use the cut levels command to specify cut ranges and the depth of the cut within each range. Cavity milling operations complete cutting at one level before moving along the tool axis to the next level.

The Cut levels dialog box is organized into separate areas: global parameters, parameters for the current range, and additional options, as shown in Fig.5-16.

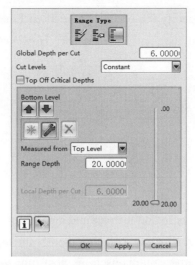

Fig.5-16 The Cut levels Dialog

Cut levels and ranges are identified in Tab.5-4.

Tab.5-4 Cut levels and ranges identified

Symbol	Description
	This symbol represents range tops, range bottoms, and critical depths. The planes are defined by geometry and are associative
	This symbol represents cut depths within a range. The planes are not associative

The active range is displayed in the color set by the Selection Visualization preference. To find the preference, choose preferences→visualization and click the Color tab. Inactive ranges are displayed in the color set by the part manufacturing geometry color preference, and top off depths are displayed in the color set by the manufacturing top off level color. To find the preference, choose preferences→manufacturing and click the Geometry tab.

You can reposition the cut level and range stack to make its display more visible. Drag the handle to a location you want.

1. Range Type

· Automatic (Default): Sets ranges to align with planar faces normal to the fixed tool axis (see Fig.5-17). Ranges define critical depths and are associative to the part. A large plane symbol with a solid outline is displayed for each range.

Fig.5-17 Automatic range

125

· User-defined: Lets you specify the bottom plane of each new range (see Fig.5-18). Ranges defined by selecting faces remain associative to the part. However, critical depths of the pare are not automatically detected.

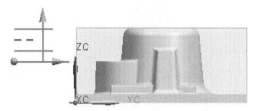

Fig.5-18 User-defined range

· Single: Sets one cut range based on the part and blank geometry (see Fig.5-19).

With the Single range type option , you can only modify the top and bottom levels of the range. If you modify either level, the same value is used the next time you process the operation. If you use the default values, they remain associative to the part. And it allows you to select top off critical depths to ensure that horizontal surfaces are added to the cut depths. The critical depths are then displayed with large plane symbols. What's more, you can specify a depth per cut value to subdivide the single range.

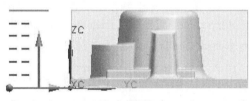

Fig.5-19 Single range

2. Cutting levels

Specifies how to subdivide the cut levels.

· Constant: Maintains a consistent depth of cut at the common depth per cut value.

· Only at Range Bottom: Does not subdivide the cut ranges.

3. Common depth per cut

Determines how to measure the default cut depth value.

· Constant: Limits the distance between successive cut passes.

· Scallop: Limits the material height between passes.

When you specify the value of common depth per cut, the software calculates equal cut levels that do not exceed the specified value. Fig.5-20 illustrates how UG NX adjusts for a value of 0.25.

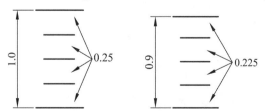

Fig.5-20 How UG NX adjusts for a value

4. Distance

Specifies the default maximum cut depth for all ranges.

5. Top off critical depths

Available when range type is set to single. Adds a cut depth for each horizontal surface (critical depth) in a part as shown in Fig.5-21.

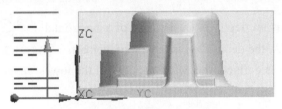

Fig.5-21   Top off critical depths

6. Top of range 1

Lets you specify the location for the top of the range.

7. Range Definition

Lets you specify parameters for the currently selected range.

· Select object: Lets you specify the location for the bottom of a range.

· Range depth: Specifies the bottom of a range. The distance is measured from the specified reference plane.

· Measured from: Specifies the reference plane (see Tab.5-5) from which to measure the range depth values. It does not affect the range definition when you select points or faces to add or modify ranges.

Tab.5-5   Reference plane specifying

Option	Description
Top Level	Measures the range depth from the top of the first cut range.
Current range top	Measures the range depth from the top of the currently highlighted range.
Current range bottom	Measures the range depth from the bottom of the currently highlighted range.
WCS origin	Measures the range depth from the origin of the WCS.

· Depth per cut: Specifies the maximum cut depth for the current active range. You can specify a different value for each cut range.

· Add new set: Adds a new range below the current active range.

· List: Displays each cut range as one row in a table that provides range depth and depth per cut information.

· The row you select in the table is the active range.

8. Cut Below Last Range

· Distance: Specifies how far to cut below the last range.

The software determines the highest and lowest cut ranges differently in cavity milling (see Fig.5-22).

(1) The default top of the highest range is the top of the part, blank, or cut area geometry.

· If cut area is defined without a blank, the default top is the top of the cut area.

· If the cut area has no depth, as in the case of a horizontal face, and there is no blank, the default top is the top of the part.

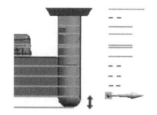

Fig.5-22 Specifies distance to cut below the last range

(2) When a cut area is defined, the default bottom of the lowest range is the bottom of the cut area.

(3) When a cut area is not defined, the default bottom of the lowest range is the bottom of either the part or blank geometry, whichever is lower.

5.4.2 Add a user-defined cut range to a cavity milling operation

The following steps show how to add a user-defined cut range to a Cavity Milling operation.

(1) Add a new cut level to an existing cavity milling operation.

· In the Cavity milling dialog box, click Cut levels 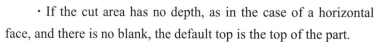.

Fig.5-23 shows a Single Range Type that spans from the bottom plane located at the bottom of the blank stock, up to the top plane located at the top of the blank stock.

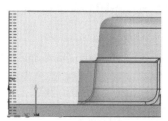

Fig.5-23 Single range type that spans from the bottom plane

· In the Cut levels dialog box, in the ranges group, from the range type list, select user-defined.

· In the range definition group, click Add new set . The new range appears in the list box.

· In the range definition group, click Select object .

· Select the part face as show in Fig.5-24.

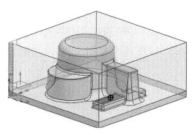

Fig.5-24 Select the part face

· In the depth per cut box (see Fig.5-25), type 0.20, and then press Enter.

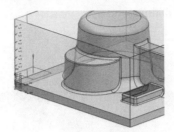

Fig.5-25　Input the depth per cut

In the range that was added:

· The bottom plane of the new range is positioned at the base of the edge that was selected.

· The new range spans from the bottom plane up to the bottom of the next range above it. If there are no ranges above, the new range spans to the top level.

· The new range is associative to the part face that was selected.

(2) Create an additional range.

· In the Cut levels dialog box, in the range definition group, click Add new set.

· In the range definition group, click Select object.

· Select the part edge as shown in Fig.5-26.

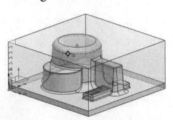

Fig.5-26　Select the part edge

· In the Depth per cut box (see Fig.5-27), type 0.050, and then press Enter.

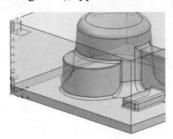

Fig.5-27　Input the depth per cut

· Click OK. The Cavity mill dialog box opens.

(3) In the actions group, click Generate.

(4) Verify the tool path.

· Click Verify. The Tool path visualization dialog box opens.

· Click the 2D dynamic tab.

- Click Play ▶ to verify the tool path (see Fig.5-28).
- Click OK. The Cavity mill dialog opens.

Fig.5-28 Verify the tool path

(5) Click OK to accept the tool path and close the dialog box.

5.4.3 Edit a range to change the cut levels

When you edit the range depth, cut levels are added or subtracted from the range. You can move the bottom plane of a cut range to any position within the cut volume in a number of ways.
- Drag the handle to change the range depth.
- Type a range depth value either in the Cut levels dialog box, or in the On-screen input box.

You can specify how far the tool can cut below the last range by entering a distance value based on the percent of the tool diameter. Cut levels before and after the range is modified are shown in Fig.5-29 and Fig.5-30 respectively.

   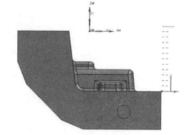

Fig.5-29 Cut levels before the range is modified   Fig.5-30 Cut levels after the range is modified

The following steps show how to edit a range to change the cut levels (see Fig.5-31).

(1) In the Cavity milling dialog box, click Cut levels 📝.

(2) In the Cut levels dialog box, the range definition group, in the list box, select a range. The range you select becomes the currently active range and is displayed in the graphics window in the selection color. Only one range can be active at a time.

(3) Change the range depth in one of the following ways.

- In the Cut levels dialog box, in the range definition group, in the range depth box, type 1.00.
- In the on-screen input box, type 1.00.
- In the graphics window, drag the range handle.

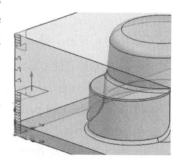

Fig.5-31 Edit a range

(4) In the Cut levels dialog box, in the range definition group, in the depth per cut box, type 0.250.

(5) Click OK. The Cavity mill dialog box opens.

(6) In the actions group, click Generate ![icon].

(7) Click OK to accept the tool path and close the dialog box.

5.4.4 Strategy tab—Extend path options

Extend at edges: it requires cut area geometry, and makes the cutter go beyond the exterior edges of the cut area to machine excess material around the part. You also can use this option to add cutting moves to the start and end of tool path passes to ensure that the tool smoothly enters and exits the part. The comparison between no extension of edges and with extension at edges are shown in Fig.5-32 and Fig.5-33.

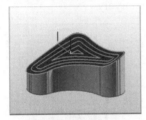
Fig.5-32   No extension of edges

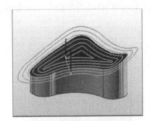
Fig.5-33   Extend at edges

5.4.5 Containment tab—IPW

Use to visualize the material (rest material) remaining from prior operations, define blank material, and check for tool collisions. There are several IPW containment options in cavity milling, as shown in Fig.5-34.

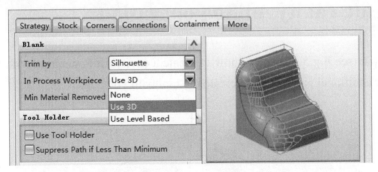
Fig.5-34   IPW options in cavity milling

· None Either uses existing blank geometry, if available, or cuts the entire cavity.

· Use level-based creates 2D representations of uncut regions for cavity milling operations.

· Use 3D creates a facet body to represent the IPW.

5.4.6 More tab—tolerant machining

Finds the correct machinable regions without gouging the part. Tolerant machining is the preferred method for most milling operations. With tolerant machining:

(1) The selected faces defining part, blank, and check geometry should enclose a volume.

(2) The material side is based on the tool axis only. The tool positions to:
 • Both sides of curves when curves are selected.
 • Both sides of vertical walls when the top face is not selected.

(3) The tool position attribute is always Tanto.

(4) The tolerance used to trace the Blank is much looser than the tolerance used to trace the Part. If the blank/geometry is very close to the size of the part geometry, the blank traces and part traces can overlap each other and result in undesirable cut regions.

(5) The software does not try to handle backdraft conditions Fig.5-35 and Fig.5-36 shows backdraft condition with tolerant machining on and off respectively.

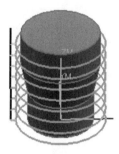
Fig.5-35　Backdraft condition with tolerant machining on

Fig.5-36　Backdraft condition with tolerant machining off

5.5　Create cavity milling operation

This example (see Fig.5-37) shows how to create a cavity mill operation with minimal inputs. You can specify additional parameters to create the tool path you need.

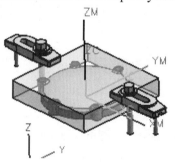
Fig.5-37　Part to be machined

132

(1) On the insert toolbar, click Create operation, or choose insert→operation.
(2) In the Create operation dialog box, in the operation subtype group, click CAVITY_MILL.
(3) In the Location group, set the options as shown in Tab.5-6.

Tab.5-6   Operation options

Program	PROGRAM
Tool	MILL
Geometry	WORKPIECE
Method	MILL_ROUGH

(4) Click OK.
(5) In the Cavity milling dialog box, in the actions group, click Generate.
(6) To view the tool path, click Verify.
(7) In the Tool path visualization dialog box, click the 2D dynamic tab.
(8) Click Play.
(9) Click OK to close the Tool path visualization dialog box.
(10) Click OK to save the operation and close the cavity milling dialog box. Cavity milling part is shown in Fig.5-38.

Fig.5-38   Cavity milling part

5.6 Cavity Milling Operation Example

This example shows how to create a cavity mill operation with additional parameters and an IPW. The part to be machined is shown in Fig.5-39.

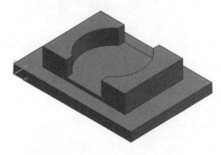

Fig.5-39   Cavity milling part model

5.6.1 Create cavity milling operation with rough method

(1) Set up the machining environment.
① Choose start→manufacturing.

② In the Machining environment dialog box, in the CAM session configuration group, select CAM general.

③ In the CAM setup to create group, select mill contour.

④ Click OK.

(2) Create machine Tool.

① On the insert toolbar, click Create tool, or choose insert→tool.

② In the Create tool dialog box, from the type list, select mill contour. In the tool subtype group, click Mill. Set machine tool name as Mill_D18.

③ The software displays the Five parameter mill Tool dialog box.

④ In the Five parameter mill tool dialog box, set the tool parameters as shown in Fig.5-40.

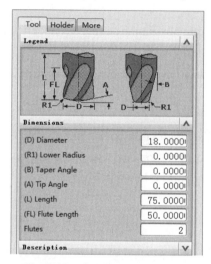

Fig.5-40 Create machine Tool

(3) Setting geometry

① In the geometry order view, double click MCS mill, the Mill orient dialog box appears.

② In the Mill orient dialog box, in the MCS group, click CSYS dialog box.

③ In the CSYS dialog box, click Dynamic.

④ In the graphics window, use the handles on the MCS to move it to the end point of the bottom surface, as shown in Fig.5-41.

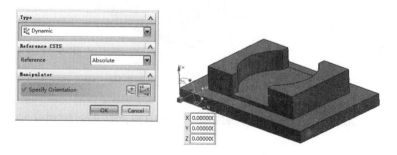

Fig.5-41 Set MCS

134

⑤ Click OK to return to the MCS dialog box.
⑥ Click OK.
⑦ In the geometry order view, double click Workpiece , the Mill geom dialog box appears.
⑧ In the Mill geom dialog box, click Specify part . The Part geometry dialog box opens.
⑨ In the graphics window, select the entire model, as shown in Fig.5-42.
⑩ Click OK to close the Part geometry dialog box.

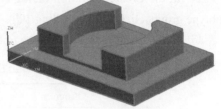

Fig.5-42 Select part geometry

⑪ In the Mill geom dialog box, click Specify blank . The Blank geometry dialog box opens.
⑫ In the selection options group , select auto block radio box. The NX creates a block containing the model automatically, as shown in Fig.5-43.

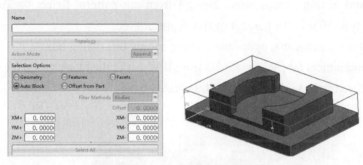

Fig.5-43 Specify blank geometry

⑬ Click OK to close the Blank Geometry dialog box.
⑭ Click OK. The workpiece has been specified.
(4) Setting machining method.
① In the machining method order view, double click Mill rough. The Mill method dialog box opens.
② In the Mill method dialog box, set parameters as shown in Fig.5-44.

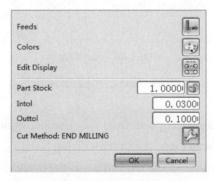

Fig.5-44 Set mill method

③ Click OK.

(5) Create operation

① On the insert toolbar, click create operation or choose insert→operation.

② In the Create operation dialog box, select mill contour from the type list. In the operation subtype group, click Cavity milling .

③ In the Location group, set the options as shown in Tab.5-7.

Tab.5-7   Operation options

Program	program
Tool	Mill D18
Geometry	workpiece
Method	Mill rough

④ Click OK. The Cavity milling dialog box is displayed.

⑤ In the path setting group, select zig-zag from cut pattern. Select % Tool flat from the stepover list, and type 50.0 in the percent of flat diameter box .

⑥ Click cutting parameters, opens the Cutting parameters dialog box.

⑦ In the containment tab, select use 3D from IPW list in the blank group (see Fig.5-45).

⑧ Click OK to return to the Cavity milling dialog box.

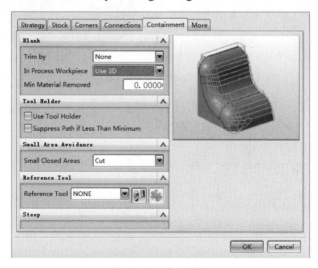

Fig.5-45   Set IPW

⑨ In the Path Setting group, Click Non-cutting moves , opens the Non-cutting moves dialog box.

⑩ In the transfer/rapid tab, select plane from the clearance option as shown Fig.5-46.

⑪ Click specify plane, opens Plane constructor dialog box.

⑫ In the graphics window, select the top surface of the model.

⑬ Type 3.0 in the Offset box.

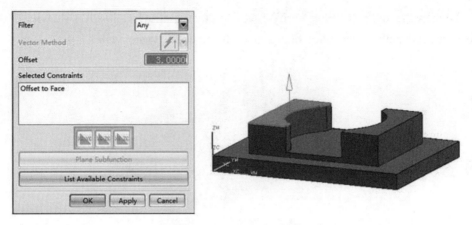

Fig.5-46 Set clearance plane

⑭ Click OK.

⑮ Click OK to close the Non-cutting moves dialog box.

⑯ In the path setting group, click Feeds and speeds , opens Feeds and speeds Dialog Box.

⑰ Set feeds and speeds parameters as shown in Fig.5-47.

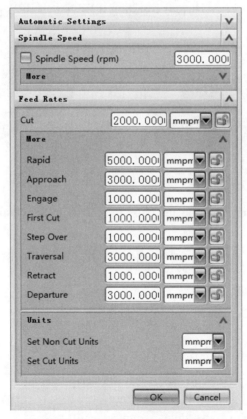

Fig.5-47 Set feeds and speeds

⑱ Click OK to return to the Cavity milling dialog box.

⑲ In the Cavity Milling dialog box, in the actions group, click Generate .

137

⑳ To view the tool path, click Verify [icon]. The cavity milling tool path is shown as Fig.5-48.

㉑ In the tool path visualization dialog box, click the 2D dynamic tab.

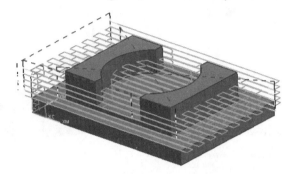

Fig.5-48　Cavity milling tool path

㉒ Select Medium from Generate IPW list, and select save IPW as component check box. The NX will add a reference set to the assembly and create a component part (see Fig.5-49).

㉓ Click Play [▶].

㉔ Click OK to close the Tool path visualization dialog box.

㉕ In the actions group, click Display resulting IPW [icon], the IPW is shown as Fig.5-50.

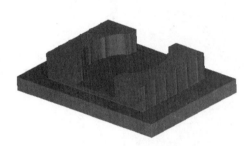
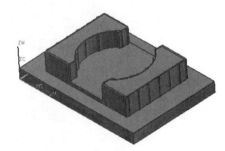

Fig.5-49　Component part　　　　　　Fig.5-50　Display resulting IPW

㉖ Click OK to save the operation and close the Cavity milling dialog box.

5.6.2　Create cavity milling operation with semi-finish method

1. Create machine Tool

① On the insert toolbar, click create tool [icon], or choose Insert→Tool.

② In the Create tool dialog box, from the type list, select mill contour. In the tool subtype group, click Mill [icon]. Set machine tool name as Mill_D12.

③ The software displays the Five parameter mill tool dialog box.

④ In the Five parameter mill tool dialog box, set the tool parameters as shown in Fig.5-51.

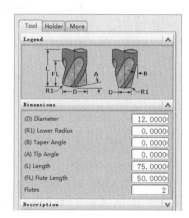

Fig.5-51　Create machine Tool

138

2. Copy operation

① In the program order view, copy the Cavity mill operation which is previously created, and paste to the PROGRAM node.

② Rename it as cavity_milling_semi-finish.

③ Double click the cavity_milling_semi-finish, The Cavity milling dialog box appears.

④ in the geometry group, click Display [icon] next to specify previous IPW.

⑤ The IPW created by the rough operation is displayed, as shown in Fig.5-52.

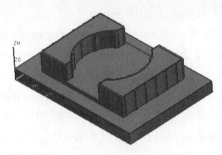

Fig.5-52　Display the previous IPW

3. Set operation parameters

① In the tool group, select mill_d12 from Tool list.

② In the path setting group, select mill_semi_finish from method list.

③ Select follow part from cut pattern list.

④ Click OK.

⑤ In the actions group, click Generate [icon]. The tool path is shown as Fig.5-53.

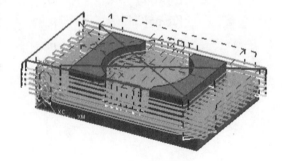

Fig.5-53　Cavity Milling Tool Path with semi-finish method

 Notes

[1] According to the order of the operations in the program order view, each operation tool path progressively reduces the IPW to mimic the material removed from the actual workpiece on the machine tool.

按照在程序命令视图中的操作顺序，每个操作的刀具轨迹使IPW逐渐减小，从而模拟在机床中从真实的工件去除材料的过程。

[2] For tool path generation, some of the operations can use the IPW shape from the previous

operation as input, which reduces user interaction and improves tool path cutting efficiency.

对于刀具轨迹的生成，一些操作可以使用前期操作的 IPW 形状作为输入，这可以减少用户的交互，并改善刀具轨迹的切削效率。

[3] In cavity milling operations, the In Process Workpiece options control what is input as the initial blank from prior operations within the same geometry group. They also control the output, which represents the state of the workpiece after completing the operation.

在型腔铣削操作过程中，IPW 选项控制着之前操作中相同几何组的原始毛坯输入，同时也控制着在操作完成后表征工件状态的输出。

[4] When you select the IPW options use 3D option or use level-based option in a cavity milling or plunge milling dialog box, and generate the operation, the software searches the geometry view hierarchy in the Operation Navigator to ensure that it meets the requirements of a valid IPW. The requirements for a valid In Process Workpiece are met.

当你在型腔铣削或插铣对话框中选择 IPW 的"使用 3D"或者"基于层"的选项，并生成操作时，软件将会在操作导航器中自动搜索几何视图层，确保能够找到有效的 IPW。

New Words in Chapter 5

mimic	['mɪmɪk]	vt.	模仿，模拟
rest mill			剩余铣
violation	[vaɪə'leɪʃn]	n.	妨碍
isolate	['aɪsəleɪt]	adj.	隔离的，孤立的
enclose	[ɪn'kləʊz]	vt.	围绕
taper	['teɪpə]	n.	锥形物
hierarchy	['haɪərɑːkɪ]	n.	层级
backdraft	['bækˌdræft]	n.	倒转，回程

Chapter 6
Face Milling

Objectives:
✓ To understand the face milling and Mixed cut pattern.
✓ To be able to create a face milling operation.

6.1 Face milling overview

6.1.1 What's face milling

Face milling operations is used to machine planar faces (see Fig.6-1). Part geometry and cut area geometry must be selected to specify the faces to be machined. Face milling removes material in planar levels with respect to the tool axis, so the faces that are not planar and perpendicular to the tool axis are ignored. For each selected cut area or boundary to be cut, traces are created from geometry, regions are identified and then cut without gouging the part.

Face milling may be used for pocketing within certain limitations.

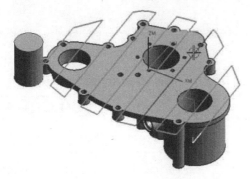

Fig.6-1   Face milling

In all face milling operations you can do the following:
· Simplify the shape of a cut pattern.
· Apply different cut patterns on multiple faces using the mixed cut pattern.
· Create a custom cut pattern using the manual cut pattern.

141

- Control how far the cutter cuts over the edge of the part.
- Apply wall stock independent of part stock.
- Extend the tool path to the part outline.
- Smooth out the tool path.
- Merge two tool paths.

6.1.2 Face milling operation subtypes

Operation subtypes (shown as Tab.6-1) that use the face milling processor cut planar faces on solid bodies, such as pads on a casting.

Tab.6-1 Face milling operation subtypes

Icon	Subtype	Description
	Face milling	The main face milling operation subtype
	Face milling area	Customized to recognize a cut area and wall selection
	Face milling area	Customized with the mixed cut pattern to let you specify a different cut pattern for each face. The available cut patterns include a manual option that lets you position the tool exactly where you need to, similar to teach mode in turning

6.2 Face milling geometry

6.2.1 Valid geometry for face milling

The following types (shown as Tab.6-2) of geometry can be specified for face milling operations.

Tab.6-2 Valid geometry for face milling

Icon	Geometry	Description
	Part	Specifies bodies that represent the finished part
	Face boundaries	Specify the faces to be machined. Available for face milling only
	Check	Specifies geometry that you want the tool to avoid
	Check boundaries	In addition to the specified part geometry, check boundaries define areas the tool must avoid

(1) Part: Solid part geometry must be specified (or inherited) in order to use gouge checking.

(2) Face boundaries: Face boundaries consist of closed boundaries whose inside material indicates the areas to be machined. Any of the following can be selected:

· A planar face, or flat B-surface

When a face boundary is created from a face, the body associated to the selected face boundary, by default, is automatically used as part geometry which is used to determine the cut regions at each level.

· Face curves and edges

· Points (which connect to form a polygon).

All members of a face boundary have tanto tool positions. At least one face boundary must be selected to generate a tool path. The normal of a face boundary plane must be parallel with the tool axis.

6.2.2 Valid geometry for face milling area

The following types (shown as Tab.6-3) of geometry can be specified for face milling area operations.

Tab. 6-3 Valid geometry for face milling area

Icon	Geometry	Description
	Part	Specify bodies that represent the finished part
	Cut area	Specify the area of the part to be machined. Available for face milling area and Face Milling only
	Wall	Specify the wall faces surrounding the cut area. Available for face milling area and face milling manual only, not available when the automatic walls check box is selected
	Check	Specify geometry that you want the tool to avoid

(1) Cut area: Cut area is an alternate method to face geometry to define the faces to be cut. to use cut area, you must not select face geometry or inherit face geometry from a mill area geometry group.

Many faces can be selected. Only the flat faces normal to the tool axis are processed. Cut area geometry can be used when:

· The face geometry (blank boundaries) is not sufficient to define the machined faces on a part body.

· Wall geometry is to be used. For example, to machine faces that have finished walls requiring a unique stock other than part stock.

(2) Wall: Use wall stock and wall geometry to override the global part stock for walls related to the machined faces on a part body.

143

6.2.3 Wall recognition in face milling area

1. Wall stock and wall geometry recognition

In a face mill area operation, you can use Specify Wall Geometry and wall stock to:
· Override the global part stock value.
· Select faces on the part body (other than the faces being machined) as wall geometry.
· Apply a unique wall stock value to those faces in place of part stock.

Fig.6-2 illustrates the different geometry types associated with the part body for a face mill area operation. In this example:
· Orange represents the face to be machined.
· Yellow represents the faces selected as wall geometry to receive wall stock.
· Green represents the faces to be treated as part geometry to receive part stock.

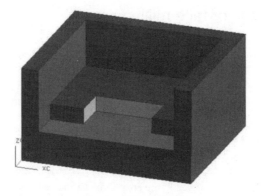

Fig.6-2 Part requiring unique wall stock

The diagram in Fig.6-3 illustrates the top view of this part when you apply a wall stock of one and a part stock of ten to the wall geometry and part geometry respectively. Notice that the system applies a wall stock of one only to the wall faces while it applies a part stock of ten to the remaining faces on the part body.

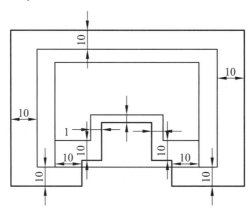

Fig.6-3 Wall stock applied to the wall faces, and part stock applied to the remaining faces on the part body

2. Using wall stock and wall geometry

Specify wall geometry [icon] is available inside the Face mill area dialog box. To use specify wall geometry, you must first select or inherit specify cut area to define the machined faces in the face milling area operation.

Wall stock may be defined in the Cutting parameters dialog box for a face milling operation.

6.3 Face milling operation parameters

6.3.1 Mixed cut pattern

The mixed cut pattern is used to specify different cut patterns on each face of the part, or when you want to use manual cut pattern.

When selecting the mixed cut pattern, different cut patterns can be used to machine each defined region of the part. Depending on the geometry of the region you are machining, you can do one of the following:

· Use a predetermined cut pattern.
· Create your own cut pattern using the manual cut pattern.
· Select omit when no cutting is needed for the current range.

Tab.6-4 is used to determine what parameters are available for specific cut methods.

Tab.6-4 Mixed cut cutting parameters

Automatic mixed cut methods	Available parameters
Zig-zag	Stepover/Cut angle/Display cut direction/Wall cleanup/Across voids/ Minimize number of engages
Zig	Stepover/Cut angle/Display cut direction/Wall cleanup/Across voids
Zig with contour	Stepover/Cut angle/Display cut direction/Across voids
Follow periphery	Stepover/Direction/Region connection/boundary approximation
Profile	Stepover/Open passes/Region connection/Boundary approximation/ Island cleanup/Additional passes
Follow part	Stepover/Open passes/Region connection/Boundary approximation/ Follow check geometry/Max traverse distance

When you select the mixed cut pattern in a face milling operation, different cut patterns can be defined in the Region cut patterns dialog box when you click Generate in the action group, as shown in Fig.6-4.

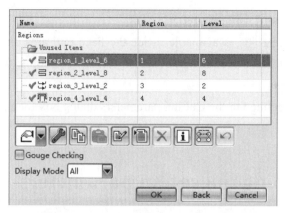

Fig.6-4   Region cut patterns dialog box

From the Region cut patterns dialog box you can:
· Assign different cut patterns to different regions.
· Edit a manual cut pattern.
· Copy a manual cut pattern from one region to another.

In the Region cut patterns dialog box, all identified regions are listed in the region column, and the different levels associated with the regions are listed in the level column. The tool cuts the entire region in one cut, then moves to the next region.

The gouge checking option prevents the tool from violating the part geometry and check geometry. The default value is on. If the tool contains a holder definition, the gouge checking is performed with the tool and holder. If the tool gouges, a warning message is displayed.

The tool path of a face milling operation with the mixed cut pattern is shown in Fig.6-5.

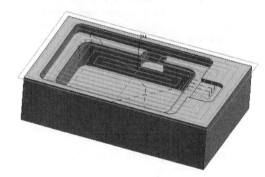

Fig.6-5   Tool path of a face milling operation with the mixed cut pattern

6.3.2   Face milling cut depth

The cut depth of each cut level is determined by three parameters: blank distance H, depth per cut h and floor stock f.

Depth per cut lets you specify the maximum depth for a cut level. It can be set in the Depth Per Cut box of Path Setting group, as shown in Fig.6-6.

Fig.6-6 Set depth per cut value in the Face milling dialog box

The actual depth will be as close to the depth per cut value as possible without exceeding it. In face milling, the cut levels N are calculated for each selected face as follows:

$$N = (H - f) / h$$

If the specified depth per cut h does not divide evenly into the material to remove, the number of cuts N is rounded up. The cut depth is then recalculated to remove the stock in even increments, as shown in Fig.6-7.

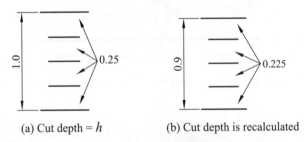

(a) Cut depth = h (b) Cut depth is recalculated

Fig.6-7 Same depth per cut value with different material thicknesses

6.4 Face milling cutting parameters

6.4.1 Blank overhang

Blank overhang is used to control the distance that the cutting tool travels beyond the edge of a face in a face milling operation. The blank overhang distance is specified as a percentage of the tool diameter. If the blank overhang value is less than 100% of the tool diameter, the tool may not be able to completely machine some areas around the convex portions of the face for zig or zig-zag cut patterns.

Fig.6-8 shows two tool paths:

· The first tool path cuts with a 100% blank overhang distance. The path extends in the cut direction, one tool diameter past the part geometry.

· The second tool path cuts with a 50% blank overhang distance. The path extends in the cut direction, half of the tool diameter past the part geometry.

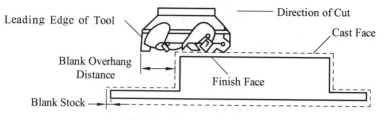

Fig.6-8   Blank overhang

Blank overhang can be set in the Blank overhang box in the Strategy tab of Cutting parameters dialog box, as shown in Fig.6-9.

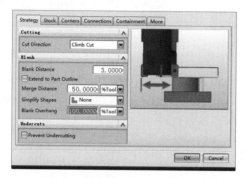

Fig.6-9   Set blank overhang value

The tool path of a face milling operation with blank overhang is shown in Fig.6-10.

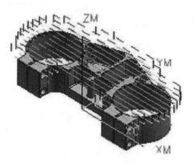

Fig.6-10   Tool path with blank overhang

6.4.2   Prevent undercutting

While face milling is designed primarily to machine planar faces, you can also use face milling to machine pockets or channels.

There may be some material that is left uncut in cases where the pocket or channel walls have a taper or blend between the walls and floor of the pocket or channel. In these cases, the prevent undercutting option can be used to control the amount of material that is left uncut. To minimize the amount of uncut material, turn prevent undercutting off as shown in Fig.6-11.

Prevent undercutting option can be turn on in the prevent undercutting checkbox in the strategy tab of Cutting parameters dialog box, as shown in Fig.6-12.

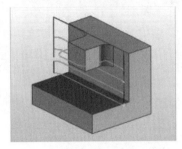 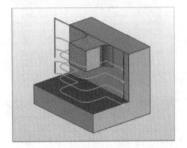

Fig.6-11 Prevent undercutting off   Fig.6-12 Prevent undercutting on

(1) Pocket with planar walls that are perpendicular to the floor.

This condition is handled the same regardless of the prevent undercutting option, as shown in Fig.6-13.

Fig.6-13 Pocket with planar walls

(2) Pocket walls with a taper.

If prevent undercutting is on (see Fig.6-14), the cutter does not move to the right beyond the dotted line. Therefore, uncut material may remain. If prevent undercutting is off (see Fig.6-15), the tool cuts to the point where it first contacts the wall.

Fig.6-14 Prevent Undercutting on   Fig.6-15 Prevent Undercutting off

(3) Blend between the wall and floor of the pocket.

If prevent undercutting is on, the cutter does not move to the right beyond the dotted line (see Fig.6-16). The material extending from the blend to the wall is left uncut. If prevent undercutting is off, the tool cuts up to where the cutter's lower radius contacts the wall (see Fig.6-17).

149

 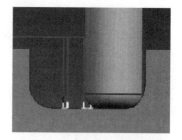

Fig.6-16 Prevent undercutting on   Fig.6-17 Prevent undercutting off

6.4.3 Merge distance

When there are multiple faces, which lie at the same level, face milling automatically attempts to merge the faces into a single region prior to machining. However, if you set the region sequencing option to standard, faces are not merged, in order to preserve the sequence in which you selected the faces.

Merge distance option can be used when you want to merge faces to be machined.

The criteria used to merge the faces is dependent upon the blank overhang percentage that you use:

• If the blank overhang percentage is greater than or equal to 50%, face milling automatically merges any faces that are within a distance of 2 × (tool diameter + horizontal clearance).

• If the blank overhang percentage is less than 50%, face milling automatically merges any faces that are within a distance of 2 × blank overhang. This is done to ensure that once faces are merged, they are likely to remain merged after the blank overhang is applied.

• It also should be noted that even when some faces get merged based on the above criteria, if there is wall geometry on the part that is not as thick as the merging criteria, the merged region may still be separated again in order to prevent the tool from gouging the wall.

Fig.6-18 shows that this part has two faces with different widths--one narrow, and one wide. The system alerts you that the blank overhang percent must be less than or equal to 100. However, be aware that your results vary greatly from a blank overhang percent of less than 100, to that of exactly 100.

Fig.6-19 shows the result of a blank overhang percent of 50.

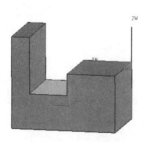 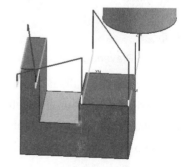

Fig.6-18 Part to be face milled   Fig.6-19 The result of a blank overhang

The preceding graphic shows, by switching the Blank Overhang percent from 50 to 100, you eliminate one of the machining passes. This is more efficient because the thin face can be machined in just one pass.

6.4.4 Extend to part outline

It extends the selected face or faces to the profile of the part. Fig.6-20 and Fig.6-21 show the difference whether face is extended.

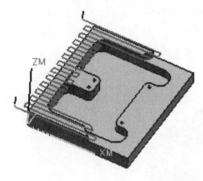 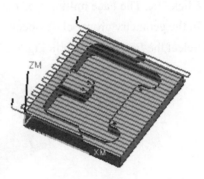

Fig.6-20 Face is not extended   Fig.6-21 Face is extended

6.4.5 Simplify shapes

Modifies complex multi-sided cut area geometry into simpler shapes. Use this option to generate a simple tool path for a complex part shape, which reduces machine motion and cutting time (see Fig.6-22, Fig.6-23, Fig.6-24).

 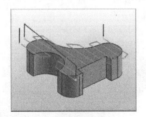 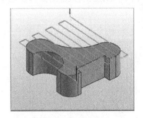

Fig.6-22 None   Fig.6-23 Convex hull   Fig.6-24 Minimum box

6.5 Create face milling operation

The following steps illustrate how to create a single level face milling operation.
(1) On the insert toolbar, click Create operation or choose insert→operation.
(2) In the Create operation dialog box, in the type group, click Mill planar.
(3) In the operation subtype group, click Face milling area.

151

(4) In the location group, select the following (see Tab.6-5):

Tab.6-5   Operation options

Program	Program
Tool	EM-.75-.03
Geometry	Workpiece
Method	Mill rough

(5) Click OK. The Face milling area dialog box is displayed.

(6) In the geometry group, click Specify cut area.

(7) Select the face (as shown in Fig.6-25) to define as your cut area.

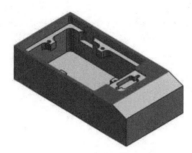

Fig.6-25   Define cut area

(8) Click OK.

(9) In the Face milling area dialog box set your parameters in the path settings group.

(10) Click Generate. The tool path is shown in Fig.6-26.

(11) Click OK to save the operation.

Fig.6-26   Tool path of a single level face milling operation

6.6   Face milling example

This example shows how to create a multi-level face milling area operation using the mixed cut pattern. The part model is shown in Fig.6-27.

Fig.6-27　Face milling operation part model

(1) On the insert toolbar, click Create operation , or choose insert→operation.

(2) In the Create operation dialog box, from the type list, select mill_planar. In the operation subtype group, click Face milling area .

(3) On the insert toolbar, click Create tool , or choose insert→tool.

(4) In the Create tool dialog box, from the type list, select mill_contour. In the tool subtype group, click Mill . The software displays the Five parameter mill tool dialog box.

(5) In the Five parameter mill tool dialog box, set the tool parameters as shown in Fig.6-28.

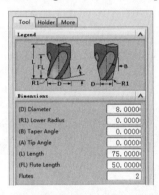

Fig.6-28　Five parameter mill tool dialog box

(6) In the Location group, set the options as shown in Tab.6-6.

Tab.6-6　Operation options

Program	Program
Tool	Mill_D8
Geometry	Workpiece
Method	Mill finish

(7) Click OK. The Face milling area dialog box is displayed.

(8) In the geometry group, click Specify cut area .

(9) In the graphics window, select multiple faces to define as your cut area as shown in Fig.6-29.

Fig.6-29　Define cut area

(10) Click OK.

(11) In the Face milling area dialog box to set your parameters in the path settings group.

(12) Click Cutting parameters .

(13) In the Cutting parameters dialog box, click the Connections tab.

(14) In the cut order group, select standard from the region sequencing list.

(15) Click OK to close the Cutting parameters dialog box.

(16) In the Face milling area dialog box, in the path settings group, from the cut pattern list, click Mixed (see Fig.6-30).

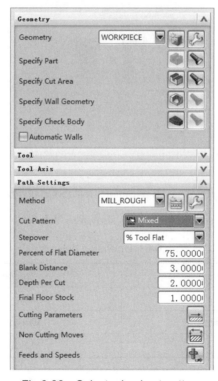

Fig.6-30　Select mixed cut pattern

(17) Click Generate .

(18) In the Region cut patterns dialog box, set a cut pattern from the list for each region (see Fig.6-31).

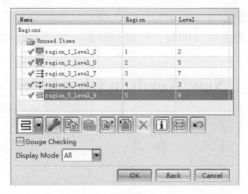

Fig.6-31　Set the cut pattern for five regions

(19) Click OK.
(20) Click Generate .
(21) Click OK to save the operation (see Fig.6-32).

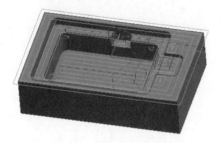

Fig.6-32　Tool path of a multiple levels face milling operation

Notes

[1] Face milling removes material in planar levels with respect to the tool axis, so the faces that are not planar and perpendicular to the tool axis are ignored. For each selected cut area or boundary to be cut, traces are created from geometry, regions are identified and then cut without gouging the part.

面铣削是在相对刀具轴的水平平面上去除材料，因此忽略了那些非平面及不垂直于刀具轴的面。对于每个选中的需要切削的区域或边界，先在几何体上创建切削轨迹，之后识别待切区域，切割没有刨削槽的部分。

[2] When a face boundary is created from a face, the body associated to the selected face boundary, by default, is automatically used as part geometry which is used to determine the cut regions at each level.

在一个面上创建面边界后，默认关联到选中的面边界的实体自动被用作工件几何体，用以决定各个层的切削区域。

[3] It also should be noted that even when some faces get merged based on the above criteria, if there is wall geometry on the part that is not as thick as the merging criteria, the merged region may still be separated again in order to prevent the tool from gouging the wall.

同样值得指出的是，即使当一些面基于以上准则发生合并时，如果在零件中壁面几何与合并的标准不一般厚，为防止刀具刨墙，已经合并的区域仍然会发生分离。

New Words in Chapter 6

merge	[mɜːdʒ]	*vi.*	合并，融合
pad	[pæd]	*n.*	衬垫
casting	[ˈkɑːstɪŋ]	*n.*	铸造，铸件
B-surface			B 级表面
polygon	[ˈpɒlɪg(ə)n]	*n.*	多边形
omit	[əˈmɪt]	*vt.*	忽略，省略

Chapter 7
Z-Level Milling

Objectives:
✓ To understand the Z-level milling and steep containment.
✓ To be able to create a Z-level milling operation.

7.1 Z-level milling overview

7.1.1 What's Z-level milling

Z-level milling is used for fixed-axis semi-finishing and finishing, which maintains a near constant scallop height and chip load on steep walls and can be especially effective for high speed machining. It removes material in planar levels that are perpendicular to a fixed tool axis (see Fig.7-1). Part geometry can be planar or contoured.

Fig.7-1 Z-level milling

With Z-level milling, you can do the following:
· Profile the entire part, or specify steep containment so that only areas with a steepness greater than the specified angle are profiled.
　· Cut multiple levels or features (regions) in one operation.
　· Cut by level (waterline) for thin-walled parts.
　· Cut an entire region without lifting the cutter.

· Maintain the tool in constant contact with the material.

7.1.2　Advantages to using Z-level

In some cases, Cavity milling with a profile cut pattern can produce a similar tool path. However, Z-level milling has the following advantages for semi-finishing and finishing.

· Z-level does not require blank geometry.

· Z-level has steep containment.

· When cutting depth first, Z-level orders by shape, where cavity milling orders by region. This means that all levels on an island part shape are cut before moving to the next island.

· On closed shapes, Z-level can move from level to level by ramping directly on the part, to create a helical-like path.

· On open shapes, Z-level can cut in alternating directions, creating a zig-zag motion down a wall.

7.1.3　Z-level milling operation subtypes

Operation subtypes that use the Z-level milling processor contour a part or cut area (see Tab.7-1). Z-level removes material in planar levels that are perpendicular to a fixed tool axis. Cutting is completed at a constant Z level before moving on to the next Z level.

Tab.7-1　Z-level milling operation subtypes

Icon	Subtype	Description
	Z-level profile	The main Z-level milling operation subtype
	Z-level corner	Customized to finish the corner areas that a previous tool could not reach due to its diameter and corner radius

7.2　Z-level milling geometry

The following types of geometry for Z-level milling operations can be specified, as shown in Tab.7-2.

Tab.7-2　Valid geometry for Z-level milling

Icon	Geometry	Description
	Part	Specifying bodies that represent the finished part
	Blank	Specifying the material to cut from

Continue

Icon	Geometry	Description
	Check	Specifying geometry to avoid
	Cut area	Specifying the areas of your part to machine
	Trim boundaries	Limiting the cut regions at each cut level

7.3 Z-level milling operation parameters

(1) Steep angle.

The steepness of the part at any given point is defined by the angle between the tool axis and the normal of the face (see Fig.7-2). When specifying steep containment, only areas with a steepness greater than the specified angle are profiled.

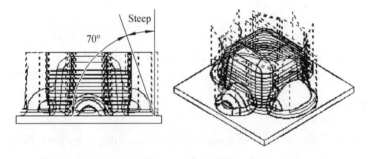

Fig.7-2 Steep angel

The options display in Z-level steepness are shown in Tab.7-3.

Tab.7-3 Z-level steepness options

Option	Description
Steep containment	Restricting the cut area based on the steepness of the part
Merge distance	Connecting the end points of cutting moves that are less than the specified distance apart to eliminate unwanted tool retracts
Minimum cut length	Eliminating tool path segments shorter than the specified value

(2) Steep containment.

Steep containment is used to control scallop height and avoid plunging the tool into the material on steep surfaces, which has two options:

· None (default): cuts steep and non-steep areas.

159

· Steep only: only cuts areas of the part with a steepness greater than or equal to the specified steep Angle.

(3) Merge distance.

Merge distance is used to eliminate unwanted tool retracts which are sometimes caused by gaps between surfaces or by small variations in steepness when the part surface steepness is very close to the specified steep angle.

(4) Minimum cut length.

Minimum cut length is used to set the minimum cut length of tool path, which can eliminate tool path segments shorter than the specified value. Tool path is not generated when the cut movement distance is smaller than the specified value of minimum cut length.

(5) Cut levels.

Cut Levels are used to specify cut ranges and the depth of the cut within each range. For Z-level milling, If a cut area is not defined, the default top of the highest range and default bottom of the lowest range are based on the top and bottom of the part geometry. If a cut area is defined, the default top of the highest range and the default bottom of the lowest range are based on the top and bottom of the cut area.

7.4 Z-level milling cutting parameters

7.4.1 Cut order

Cut order specifies how to process a cutter path with multiple regions. For cavity milling, the cutter path is ordered by cut areas. While for Z-level milling, the cutter path is ordered by the shape of cutting geometry.

7.4.2 Between levels

Between levels is available for Z-level Milling only. It allows users to cut all levels without lifting back up to the clearance plane. Four options are available for cut to cut.

(1) Use transfer method.

It lifts the cutter to the clearance plane after each pass (see Fig.7-3). The software uses the clearance information specified in the Non-cutting moves dialog box.

(2) Direct on part.

It follows the part, like a stepover move (see Fig.7-4). A cut region start point can be used to locate these moves. It's different from direct transfer, which is a rapid move with no gouge or collision checking.

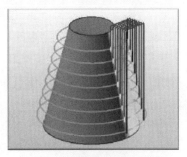
Fig.7-3 Use transfer method

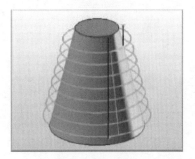
Fig.7-4 Direct on part

(3) Ramp on part.

It is not available to machine open regions (see Fig.7-5). Follows the part from one level to the next, at the ramp angle specified in the engage and retract parameters. It has a more constant depth of cut and scallop, and makes a complete pass at the top and bottom. A cut region start point can be used to locate the ramps.

(4) Stagger ramp on part.

It is not available to machine open regions (see Fig.7-6). This option is similar to the ramp on part option, except that it completes each pass before ramping to the next level.

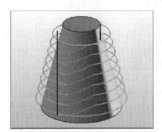
Fig.7-5 Ramp on part

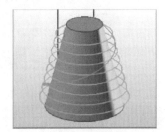
Fig.7-6 Stagger ramp on part

7.4.3 Cut between levels

Cut between levels option is used to create extra cuts when there is a gap between the cut levels in Z-level machining, which eliminates the large scallops in shallow regions left from standard level-to-level machining and eliminates the need to create a separate area milling operation for non-steep areas, or to use a very small depth of cut to control scallops in non-steep areas. It can minimizes excessive tool wear or even breakage from rapidly loading and unloading the tool in areas that have large scallops left behind from previous operations, producing a more uniform stock for semi-finishing and a more consistent surface finish with fewer retracts and engages for finishing.

· Cut Between levels off: Do not create extra cuts when there are gaps between cut levels (see Fig.7-7).

· Cut Between levels on: Creates extra cuts when there is a gap between the cut levels in Z-level machining to eliminate large scallops in shallow regions (see Fig.7-8).

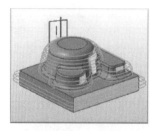
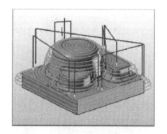

Fig.7-7　Cut between levels off　　　　Fig.7-8　Cut between levels on

Fig.7-7 shows the cut path with cut between levels off. Notice the shallow regions have large gaps. Fig.7-8 shows the cut path with cut between levels on. Notice it produces a more uniform cut path and eliminates large gaps between cut levels.

1. Top off critical depths in a Z-level operation with gaps

Topping off critical depths affects the gap regions for Z-level. Fig.7-9 shows how a single gap region is broken into two regions by the top off critical depths tool path.

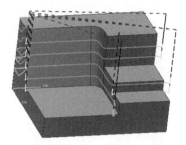

Fig.7-9　Top off critical depths tool path

To top off critical depths:

① In the operation dialog box, click Cut levels 📝.
② In the Cut levels dialog box, click Single ▮.
③ Select the Top off critical depths check box.

2. Tops and bottoms of ranges

Gap tool paths are restricted to cut area surfaces that are also inside the defined cut levels range. Surfaces are not cut when they were not selected as a cut area surface or they are outside of the Z-level range.

When the top or bottom of a range is at a horizontal surface, and the surface is selected as a cut area, the tool path is generated based on the parameters set for cut between levels. In Fig.7-10, the orange faces are the cut area. The top of the pad ① is at the top of the cut levels range, so it is cut. The floor of the pocket ② is outside of the cut levels, so it is not cut. The flat face across the entire block is within the cut levels, but not part of the orange cut area, so it is not cut.

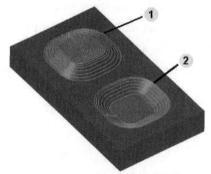

Fig.7-10　Gap tool path

3. Sequencing gap and Z-level tool paths

The Z-level and gap tool paths are sequenced and ordered as follows:

· The Z-level tool path is machined top-down and uses the same connection methods as it would without the cut between levels option.

· Level-to-level connections that violate a gap region are replaced with a traversal move.

· The start point of a new level is placed so that it isn't adjacent to a gap region. (This applies to a closed boundary.) This prevents the connection to the next level going through the gap.

· Gap machining engages and retracts from the gap paths along the tool axis. Engage/retracts to and from standard Z-level cuts use the style you specified.

· The gap tool path is machined immediately after its lower trace.

· Connections are made from the Z-level cut to the gap area. After the gap area is cut, the tool returns to the lower level.

· The lower trace is split in order to minimize retracts and engages to and from the gap tool path.

After each Z-level cut is completed, the tool begins to cut the level below it. While cutting the lower level, the system looks for gaps between lower level and the previous level above it. When a gap is found, the system cuts the gap, and then continues cutting the lower level until another gap is found, or the level is completed. The system connects the gap cutting with the lower cut level based on the max traverse distance value. If max traverse distance is exceeded, a traverse to the next level takes places. If the move to the next level is within the max traverse distance value, the tool makes a direct on-part move (shortest distance) to the next level without traversing.

4. Effect of max traverse distance

When the tool moves between gap regions, depending on the max traverse distance value, it either traverses along the part to the start point on the lower contour, or retracts and traverses to the next Z-level or gap machining area. To eliminate motions where the tool is not cutting, set max traverse distance to zero. Engages and retracts then occur directly along the tool axis.

Fig.7-11 and Fig.7-12 show how the tool moves from the lower Z-level tool path to the gap tool path area.

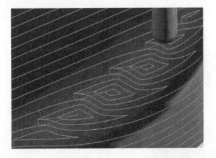

Fig.7-11 Moving from lower z-level to the gap tool path

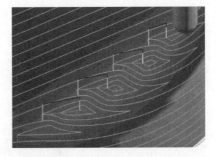

Fig.7-12 Same part with max traverse distance exceeded

5. Traversing multiple sub-regions

Fig.7-13 illustrates how the software traversed between multiple sub-regions.

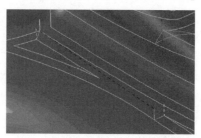

Fig.7-13 Traversing between multiple sub-regions

7.4.4 Feed on short moves

Feed on short moves specifies how to connect different cutting areas within a region, which is available when the cut between levels checkbox is on.

• Feed on short moves off: Specifies that the cutter use the current transfer method to retract, and then traverse to the next location and engage (see Fig.7-14).

• Feed on short moves on: Moves the cutter along the part surface, at the stepover feed rate, when the distance is less than the max traverse distance value (see Fig.7-15). If the distance is greater than the max traverse distance value, the cutter uses the current transfer method to retract, then traverse to the next location and engage.

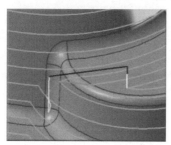 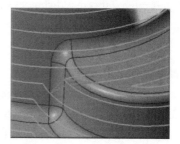

Fig.7-14 Feed on short moves off   Fig.7-15 Feed on short moves on

• Max traverse distance: Specifies the maximum allowable distance the cutter can move along the part surface at the stepover feed rate when not cutting (see Fig.7-16).

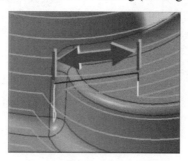

Fig.7-16 Max traverse distance

7.4.5 Reference tool

Reference tool is used to eliminate tool motion where there is no material. It references a larger tool from a previous operation. The smaller tool in the current operation removes the remaining material in the uncut regions where the larger reference tool could not fit.

Example of Z-level operations with reference tool is show in Fig.7-17.

· Overlap distance: Extends the width of the area to machine for the specified distance along the tangent surfaces of the remaining material (see Fig.7-18).

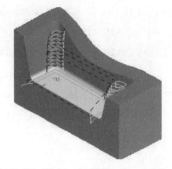

Fig.7-17 Z-level with reference tool

Overlap distance is used to help eliminate scallops and obtain a full clean up between tool paths in a sequence of rest milling operations. The applied overlap distance value is limited to the tool radius.

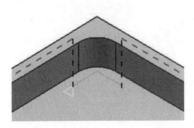 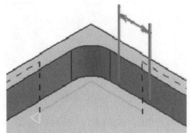

(a) Overlap distance set to zero　　　　(b) Overlap distance set to a value of 15.00

Fig.7-18 Overlap distance

Reference tool and cut between levels cannot be used together. Selecting a reference tool will deactivate cut between levels and its related parameters.

7.5 Z-level milling example

This example shows how to create Z-level milling operations. The part has been machined roughly with Cavity Milling operation, as shown in Fig.7-19.

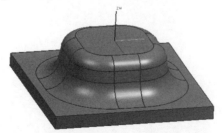

Fig.7-19 Z-level milling operation model

7.5.1 Create Z-level profile milling operation

(1) On the insert toolbar, click Create operation, or choose insert→operation.

(2) In the Create operation dialog box, from the type list, select mill contour. In the operation subtype group, click Z-level profile.

(3) On the insert toolbar, click Create tool, or choose insert→tool.

(4) In the Create tool dialog box, from the type list, select mill_contour. In the tool subtype group, click Mill. The software displays the Five parameter mill tool dialog box.

(5) In the Five parameter mill tool dialog box, set the tool parameters as shown in Fig.7-20.

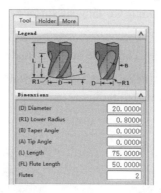

Fig.7-20 Five parameter mill tool dialog box

(6) In the location group, set the options as shown in Tab.7-4.

Tab.7-4 Operation options

Program	Program
Tool	Mill_d20
Geometry	Workpiece
Method	Mill finish

(7) Click OK. The Z-level profile dialog box appears as shown in Fig.7-21.

Fig.7-21 Z-level profile dialog box

(8) In the path setting group, type 2.0 in the global depth per cut box.

(9) In the path setting group, Click Cutting parameters ![icon]. The Cutting parameters dialog box appears as shown in Fig.7-22.

(10) In the Cutting parameters dialog box, click the Connections tab. In the level to level list, select direct on part.

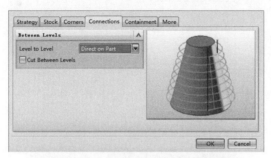

Fig.7-22　Cutting parameters dialog box

(11) In the Actions group, click Generate ![icon].

(12) Click OK to accept the tool path and close the dialog box.

In Fig.7-23, there is no cutter lifting in the tool path, which shortens the cutting feed time dramatically.

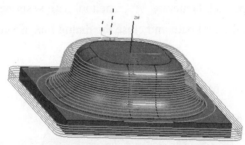

Fig.7-23　Tool path of Z-level milling with direct on part option

(13) In the path setting group, Click Cutting levels ![icon]. The Cutting levels dialog box appears.

(14) In the Cutting levels dialog box, select User-defined in the range type group.

(15) Click Delete current range repeatedly until all the cutting levels are deleted.

(16) Click Insert range, the range depth box is activated. In graphic windows, select the surfaces as shown in Fig.7-24.

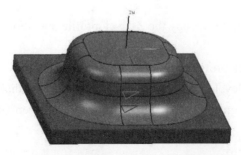

Fig.7-24　User-defined cutting levels

(17) In the local depth per cut box, type 2.0.

(18) Click OK. The Z-level profile dialog box appears as shown in Fig.7-25.

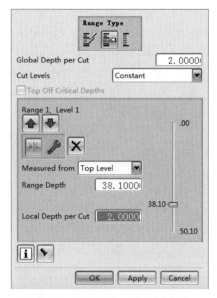

Fig.7-25 Cutting levels dialog box

(19) In the actions group, click Generate . The tool path is shown in Fig.7-26.

(20) Click OK to accept the tool path and close the dialog box as shown in Fig.7-26.

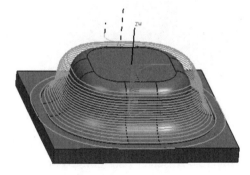

Fig.7-26 Tool path of Z-level milling with user defined cutting levels

7.5.2 Create Z-level profile milling operation with steep containment

(1) In the program order view, copy the Z-level_profile operation which is previously created, and paste to the program node.

(2) Rename it as Z-level_profile_steep.

(3) Double click the Z-level_profile_steep, The Z-level profile dialog box appears.

(4) In the path setting group, select steep only from the steep containment list as shown in Fig.7-27.

(5) Type 70.0 at angle box.

Fig.7-27 Setting steep containment in Z-level profile milling

(6) In the Actions group, click Generate . The tool path is shown in Fig.7-28.

(7) Click OK to accept the tool path and close the dialog box.

Fig.7-28 Tool path of Z-level milling with steep containment

Notes

[1] Z-level milling is used for fixed-axis semi-finishing and finishing, which maintains a near constant scallop height and chip load on steep walls and can be especially effective for high speed machining.

等高轮廓铣用以固定轴的半精加工和精加工，保持一个接近常数的残余高度和陡峭壁上的碎屑。等高轮廓铣对高速加工尤其有效。

[2] Cut Between Levels option is used to create extra cuts when there is a gap between the cut levels in Z-level machining, which eliminates the large scallops in shallow regions left from standard level-to-level machining and eliminates the need to create a separate area milling operation for non-steep areas, or to use a very small depth of cut to control scallops in non-steep areas.

当等高轮廓加工中的切削层中存在间隙时，可以用基于层的切削选项来创建额外的切削。这减少了标准的层到层加工产生的浅层区域的大量残余高度，同时也减少了为非陡峭区域创建一个独立的区域铣削操作，或者用一个非常小的切削深度来控制非陡峭区域的残留高度的必要性。

[3] Gap tool paths are restricted to cut area surfaces that are also inside the defined cut levels

range. Surfaces are not cut when they were not selected as a cut area surface or they are outside of the Z-level range.

间隙刀具轨迹被限制在切削区域表面及定义的切削层范围里面。当这些表面没有被选择为切削区域或在等高轮廓范围之外时，它们将不被切削。

New Words in Chapter 7

chip	[tʃɪp]	n.	切屑，凿
steepness	['stipnɪs]	n.	陡峭，陡坡
ramp	[ræmp]	vi.	蔓延，使有斜面
segment	['segm(ə)nt]	n.	部分，段
stagger	['stægə]	adj.	交错的，错开的

Chapter 8
Fixed-axis Surface Contouring

Objectives:
- ✓ To understand the fixed-axis surface contouring .
- ✓ To understand the drive methods of fixed-axis surface contouring.
- ✓ To be able to create a fixed-axis surface contouring operation.

8.1 Surface contouring overview

8.1.1 Surface contouring introduction

Fixed-axis surface contouring operations and variable-axis surface contouring operations finish areas on contoured surfaces by removing material along the part contours. You can control the tool axis and projection vector options to create tool paths that follow the contours of very complex surfaces.

For fixed-axis surface contouring operations (mill contour operation type), the tool axis remains parallel to a specified vector as shown in Fig.8-1. For variable-axis surface contouring operations (mill multi-axis operation type), the tool axis constantly changes orientation as it moves along the tool path, as shown in Fig.8-2.

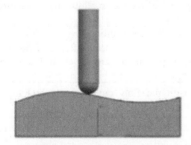 

Fig.8-1 Fixed tool axis   Fig.8-2 Variable tool axis

To create a surface contouring operation, you should specify part geometry (optional), drive geometry, projection vector (required if part geometry is specified) and tool axis.

Drive geometry can include part geometry or geometry that is not associated with the part. The software creates drive points from the specified drive geometry to control the tool position.

(1) When part geometry is not specified, the software positions the tool directly on the drive points.

(2) When part geometry is specified, the software positions the tool where the drive points project onto the part geometry.

The selected drive method determines how you define the drive points required to create a tool path. Some drive methods create a string of drive points along a curve while others create an array of drive points within an area.

The projection vector defines how the drive points project to the part surface, and which side of the Part Surface the tool contacts.

The selected drive method determines which projection vectors are available.

8.1.2 Fixed-axis surface contouring operation subtypes

Fixed-axis surface contouring removes material along the part contours. Operation subtypes that use the *fixed-axis surface contouring* processor contour a part or cut area (see Tab.8-1).

Tab.8-1 Fixed-axis surface contouring operation subtypes

Icon	Subtype	Description
	Fixed contour	The main fixed-axis surface contouring operation subtype
	Contour area	Customized with the area milling drive method to cut selected faces or a cut area. It is commonly used for semi-finishing and finishing
	Contour surface area	Customized with the surface area drive method to cut a single drive surface, or a well-ordered rectangular grid of drive surfaces
	Streamline	Customized with the streamline drive method to cut drive surfaces defined by curve sets which can be automatically generated from the part geometry
	Contour area dir steep	Customized with the area milling drive method to cut only steep areas. Use it with the contour zig-zag or contour area operation subtypes to reduce scallops by criss-crossing a previous zig-zag cut
	Contour non-steep area	Customized with the area milling drive method to cut only non-steep areas. It is often used after a Z-Level profile operation with steep containment to control scallops when finishing a cut area
	Flowcut single	Customized with the flow cut drive method to finish or relieve corners and valleys
	Flowcut multiple	Customized to cut multiple passes with the flow cut drive method
	Flowcut ref tool	Customized to cut multiple passes, based on a previous reference tool diameter, with the flow cut drive method. It is used to remove the remaining material in corners and valleys
	Contour text	Customized to cut the text in a drafting note. It is used for 3D engraving

8.2 Valid geometry for fixed-axis surface contouring

The following types of geometry (as shown in Tab.8-2) can be specified for fixed-axis surface contouring operations.

Tab.8-2 Valid geometry for fixed-axis surface contouring

Icon	Geometry	Description
	Part	Specify bodies that represent the finished part
	Check	Specify geometry that you want the tool to avoid
	Cut Area	Specify the areas of your part to machine
	Trim Boundaries	Limit the cut regions at each cut level
A	Drafting Text	Define text for engraving

The specified part geometry works in combination with the drive geometry (see Fig.8-3, usually a boundary) to define the cut region.

You may specify bodies (sheet or solid), faceted bodies, surface region features, or faces as part geometry.

Selecting a solid body provides several advantages.

· Change processing is easier, because the associativity is maintained to the entire solid body, and not the individual faces, which may change when the solid updates.

· It is easier to select a solid body than individual faces.

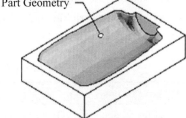

Fig.8-3 Part geometry

If you want to cut only some of the faces on a solid, you can use cut area, drive geometry, or other containment, depending on the drive method, to limit your cutting to less that the entire part.

Surface region features offer a view of part geometry and how it relates to the drive geometry.

8.3 Drive methods

8.3.1 Drive method overview

Drive methods define the drive points required to create a tool path. Some drive methods allow you to create a string of drive points along a curve while others allow you to create an array of drive points within a boundary or on selected surfaces. Once defined, the drive points are used to create a tool path. If no part geometry is selected, the tool path is created directly from the drive points. Otherwise, the drive points are projected on to the part surfaces to create the tool path.

Selecting the appropriate drive method should be determined by the shape and complexity of the surface you wish to machine, and the tool axis and projection vector requirements. The selected drive method determines the type of drive geometry you can select, as well as the available projection vectors, tool axis, and cut types.

Projection vector is an option common to most drive methods. It determines how the drive points project on to the part surfaces, and which side of the part surface the tool contacts. The available projection vector options will vary depending on the drive method used.

Fig.8-4 illustrates an example of the surface area drive method. This drive method was selected because of the complexity of the part surfaces and the required control of the tool axis. An array of drive points is created on the selected drive surface and then projected along a specified projection vector to the part surfaces. The tool positions to the part surfaces at contact points. The tool path is created using the output cutter location point at the tip of the tool. Both the projection vector and the tool axis are variable and defined as normal to the drive surface.

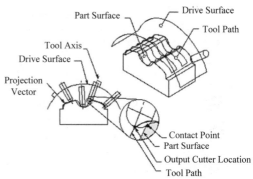

Fig.8-4 Surface area drive method

The drive methods of surface contouring are shown in Tab.8-3.

Tab.8-3 The drive methods of surface contouring

Drive methods	Description
Curve/Point	Defines drive geometry by specifying points and selecting curves
Spiral	Defines drive points that spiral outward from a specified center point
Boundary	Defines cut regions by specifying boundaries and loops
Area milling	Defines cut regions by specifying cut area geometry. Drive geometry isn't required
Surface area	Defines an array of drive points that lie on a grid of drive surfaces
Tool path	Defines drive points along the tool path of an existing CLSF to create a similar Surface Contouring tool path in the current operation
Radial cut	Generate Drive Paths perpendicular to and along a given boundary, using a specified Stepover distance, Bandwidth and Cut Type
Contour profile	Machines canted walls with the side of the cutter
Flow cut	Generates drive points along concave corners and valleys formed by part surfaces
Text	select annotation and specify a depth to engrave text on the part

8.3.2 Curve/Point drive method

1. Curve/Point drive method overview

Use the curve point drive method to define drive geometry by specifying points and selecting curves or face edges.

· When specifying points, the drive path is created as linear segments between the specified points.

· When specifying curves or edges, drive points are generated along the selected curves and edges.

The drive geometry is projected on to the part geometry where the tool path is then created. The curves may be open or closed, contiguous or non-contiguous, planar or non-planar.

For scribing type operations, part geometry can be left undefined to allow the cutter to cut into the part.

2. Drive geometry using points

When points define the drive geometry, the cutter moves along the tool path from one point to the next in the order you select (see Fig.8-5). You can use the same point more than once, provided you do not select it consecutively in the sequence. You can create a closed path by selecting the same point as the first and last point in the sequence (see Fig.8-6).

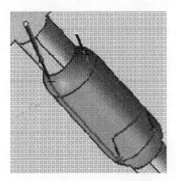   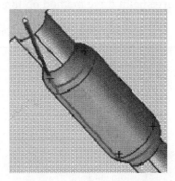

Fig.8-5 Points for open path   Fig.8-6 Points for closed path

If you specify only one drive point or several drive points such that only one position is defined on the part geometry, a tool path is not generated and an error message is displayed.

3. Drive geometry using curves and edges

When curves or edges define the drive geometry (see Fig.8-7), the cutter moves along the tool path from one curve or edge to the next in the selected order. Selected curves may be contiguous or noncontiguous.

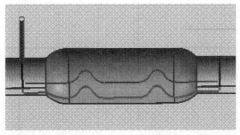

Fig.8-7 Drive geometry defined by curves

For open curves and edges, the selected end determines the start point. For closed curves or

175

edges, the start point and cut direction are determined by the order in which the segments are selected. The origin and cut direction are determined by the order of selection. The origin can be modified with the specify origin curve command.

You can use a negative stock value with this drive method to allow the tool to cut just below the selected part surface, creating a groove as shown in Fig.8-8.

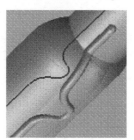

Fig.8-8　Negative stock groove

4. Curve point drive method options

The curve point drive method options are shown in Tab.8-4.

Tab.8-4　Curve point drive method options

	Items	Description
Drive geometry	Select curve	Lets you specify the curve by selecting existing curves, edges, or points. The point constructor allows points to be created and selected as stand alone flow/cross entities or to bridge gaps
	Reverse direction	Active when curves or edges are selected. Reverses the direction of an active drive set
	Specify origin curve	Lets you specify the origin when you select multiple curves or edges that form a closed loop
	Custom cut feed rate	Add custom feed rate values to each set of curves
	Add new set	Creates a new (empty) drive set in the list and activates select curve. The new set is placed in the list right after the active drive set
Drive settings	Cut step	Choose between number or tolerance
	Number	Controls the number of points placed on each curve or edge. The more points, the smoother the tool path
	Tolerance	Allows you to specify the maximum allowable chordal deviation between the drive curve and the line extending between two consecutive drive points

8.3.3　Spiral drive method

The spiral drive method enables you to define drive points that spiral outward from a specified center point (see Fig.8-9). The drive points are created within the plane normal to the projection vector and containing the center point, and then projected on to the selected part surfaces along the projection vector.

Unlike other drive methods which require an abrupt change in direction to stepover to the next cutting pass, spiral drive method stepovers are a smooth, constant transition outward. Because this drive method maintains a constant cutting speed and smooth motion, it is useful for high speed machining applications.

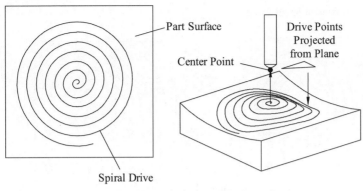

Fig.8-9 Spiral drive method

The center point defines the center of the spiral and it is where the tool begins cutting. If a center point is not specified, the system uses 0,0,0 of the ACS. If the center point is not on the part surface, it follows the defined projection vector to the part surface. The direction of the spiral (clockwise vs. counterclockwise) is controlled by the climb or conventional cut direction.

The options display in the Spiral drive method dialog box are shown in Tab.8-5.

Tab.8-5 Spiral drive method options

Items	Description
Specify point	Displays the point subfunction dialog allowing you to define the center point of the spiral drive path
Stepover	Specifies the distances between successive cut passes
Max spiral radius	Limits the area to be machined by specifying a maximum radius
Cut direction	Defines the direction the drive path cuts in relationship to the spindle rotation
Display	Creates a temporary display showing the drive path used to generate the tool path. The drive path is generated and displayed in the plane normal to the projection vector and containing the center point.

Spiral drive method stepovers are a smooth, constant transition outward and do not require an abrupt change of direction as shown in Fig.8-10.

Maximum spiral radius (see Fig.8-11) allows you to limit the area to be machined by specifying a maximum radius. This constraint reduces processing time by limiting the number of drive points created. The radius is measured in the plane normal to the projection vector.

If this specified radius is contained within the part surfaces, the center of the tool positions to the radius before retracting. If the specified radius exceeds the part surfaces, the tool continues to cut until it can no longer position to the part surface. The tool then retracts and

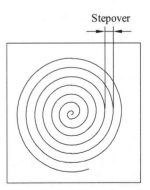

Fig.8-10 Stepover distance for spiral drive

engages (see Fig.8-12) when it can once again position to the part surface as illustrated below.

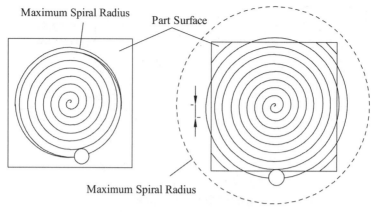

Fig.8-11   Max spiral radius

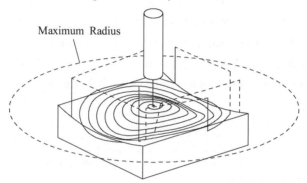

Fig.8-12   Retract and engage

8.3.4 Boundary drive method

The boundary drive method enables you to define cut regions by specifying boundaries and containment loops (see Fig.8-13). Boundaries are not dependent on the shape and size of the part surfaces while loops must correspond to exterior part surface edges. Cut regions are defined by boundaries, loops, or a combination of both. The tool path is created by projecting drive points from the defined cut region to the part surfaces in the direction of a specified projection vector. The boundary drive method is useful in machining part surfaces requiring minimal tool axis and projection vector control.

The boundary drive method works in much the same way as planar milling. Unlike planar milling, however, the boundary drive method is intended to create finishing operations that allow the tool to follow complex surface contours.

Like the surface area drive method, the boundary drive method creates an array of drive points contained within an area. Defining drive points within a boundary is generally faster and easier than selecting drive surfaces. But you cannot control the tool axis or the projection vector relative to a drive surface when using the boundary drive method. Planar boundaries, for example, cannot wrap around complex part surfaces to evenly distribute drive points or control the tool axis as illustrated in the following figure.

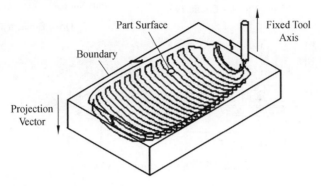

Fig.8-13 Boundary drive method

Boundaries may be created from a sequence of curves, an existing permanent boundary, points, or a face. They can define the exterior of the cut region as well as islands and pockets. Each boundary member is assigned an on, tanto, or contact tool position attribute. For a detailed discussion of how to create boundaries, refer to permanent boundaries.

A boundary may exceed the size of the part surfaces, restrict a smaller area within the part surfaces, or coincide with the edges of the part surfaces as illustrated in Fig.8-14. When the boundary exceeds the size of the part surfaces by a distance greater than the tool diameter, edge tracing may occur. This is generally an undesirable condition where the tool rolls over the edge of the part surface.

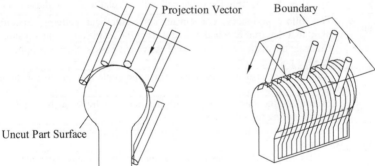

Fig.8-14 Drive points projecting to part surface

When a boundary restricts an area within the part surface, the tool must position to the boundary using on, tanto, or contact. When the cut region coincides with the exterior edge, it is best to use part containment loops (as opposed to boundaries) specified as on, tanto, or contact. These three options particularly affect how the tool positions on very steep part surfaces.

Drive boundaries contain the tool for surface contouring operations in much the same way as part boundaries do for planar and cavity milling operations. A drive boundary, however, may not fully contain the tool in a variable axis operation. In Fig.8-15, for example, the side of the tool violates the drive boundary and gouges the part while maintaining a normal tool axis orientation to the part surface. For this reason, drive boundaries should generally be used only for fixed axis operations.

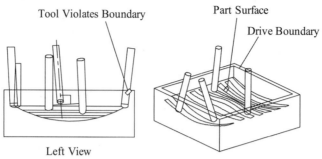

Fig.8-15　Drive boundary violated

The options display in the Boundary drive method dialog box are shown in Tab.8-6.

Tab.8-6　Boundary drive method options

Items	Description
Specify drive geometry	Lets you specify a boundary to define the drive geometry
Boundary intol/ boundary outtol	Lets you specify the inside and outside tolerance values for the boundary
Boundary offset	Lets you specify how much material to leave along the boundary
Part containment	Defines cut regions by creating loops along exterior edges of selected part surfaces and surface regions
Cut pattern	Specifies the shape of the tool path and how the cutter moves from one cut pass to the next
Cut direction	Specifies the direction the drive path cuts in relationship to the spindle rotation
Pattern direction	For follow periphery, concentric, and radial cut patterns. Specifies a pocketing method that determines whether to cut from the inside out or the outside in
Stepover	Controls the distance between successive cut passes. Stepover options vary depending on the cut pattern used
Cut angle	Automatic or user-defined. Determines the angle of rotation for the parallel lines cut patterns
Region connection	Minimizes the number of engage, retract, and traversal moves that occur between different cutting regions on a part
Boundary approximation	Reduces processing time by converting, curving, and cutting passes into longer linear segments
Island cleanup	Inserts an additional pass around islands to remove any extra material that may have been left behind
Wall cleanup	Removes the ridges that occur along the walls of a part
Finish pass	Not available for standard drive and profile cut patterns Adds a finishing cutting pass at the end of the normal cutting operation to trace around the boundaries
Finish stock	Stock value for the finish pass
Cut regions	Defines start points and specifies how cut regions are graphically displayed for visual reference

8.3.5　Area milling drive method

The area milling drive method (see Fig.8-16) enables you to define a fixed axis surface contouring operation by specifying a cut area and, if desired, adding steep containment and trim boundary constraints. It is similar to the boundary drive method, but requires no drive geometry and uses a robust and automated computation of collision-free containment. It is available only for fixed-axis surface contouring operations. For these reasons, area milling drive method should be used in place of the boundary drive method whenever possible.

Cut area may be defined by selecting surface regions, sheet bodies, or faces. Unlike the surface area drive method, the cut area geometry need not be selected in an orderly grid of rows and columns.

If a cut area is not specified, the system will use the entire defined part geometry (excluding areas not accessible by the tool) as the cut area. In other words, the system will use the silhouette of the part as the cut area. Edge tracing cannot be removed if the entire part geometry is used and no cut area is defined.

Area milling drive method may use zig-zag with lifts cut type which lifts the tool between passes according to the specification for local engage, retract, and traverse moves. It does not output departure and approach moves.

In the fixed contour dialog, the drive method area, specify the desired parameters by pressing the wrench icon, select your parameters, and press OK to accept. In the Fixed contour dialog, choose cut area to define the cut area geometry. If the cut area geometry is not defined, the system will use the silhouette of the part.

The cut regions may be further constrained by using trim. You may define the area of the cut region to be excluded from the operation by specifying trim side as inside or outside (see Fig.8-17). Trim boundaries are always closed, always use an on condition, and are projected to the part geometry along the tool axis vector. More than one trim boundary may be defined. In the Cutting dialog box, a boundary stock can be specified to define the distance the tool is positioned from the trim boundary, and boundary Intol/Outtol.

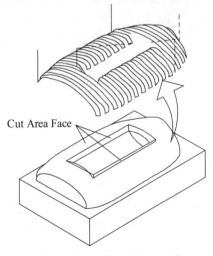

Fig.8-16 Area milling drive method

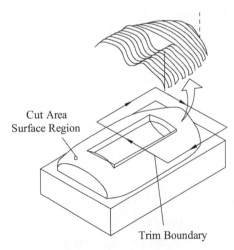

Fig.8-17 Trim boundary

The options display in the Area milling drive method dialog box are shown in Tab.8-7

Tab.8-7　Area milling drive method options

Items	Description
Method	Restricts the cut area based on the steepness of the part surfaces
Pattern	Specifies the shape of the tool path
Cut type	Specifies how the cutter moves from one cut pass to the next. (Parallel lines only)
Cut direction	Climb cut and conventional cut allow you to define the direction the drive path cuts in relationship to the spindle rotation. These options are only available for the zig, zig with contour, and zig with stepover cut types
Stepover	Controls the distance between successive cut passes. The cut pattern selected determines the available stepover options
Stepover applied	On plane or on part
Cut angle	Automatic, user-defined or longest line. Determines the angle of rotation for the parallel lines cut patterns
Additional passes	Lets you specify an additional number of passes that allow the tool to step outward in successive concentric cuts for profile cutting patterns. Refer to the discussion of additional passes in the boundary drive method for additional details
Region connection	Minimizes the number of engage, retract, and traversal moves that occur between different cutting regions on a part. Follow periphery and profile cut patterns only
Cut regions	Lets you define cut region start points and specify how cut regions are graphically displayed for visual reference

Steep containment group restricts the cut area based on the steepness of the tool path.

· None: Imposes no steepness restrictions on the tool path and machines the entire cut area.

· Non-steep: Only machines within the cut area where the part surface angle is less than the steep angle value.

· Directional steep: Only machines within the cut area where the part surface angle is more than the steep angle value.

8.3.6　Surface area drive method

The surface area drive method enables you to create an array of drive points that lie on a grid of drive surfaces. This drive method is useful in machining very complex surfaces requiring a variable tool axis. It provides additional control of both the tool axis and the projection vector.

In this example (shown as Fig.8-18), both the projection vector and the tool axis are variable and are defined as normal to the drive surface.

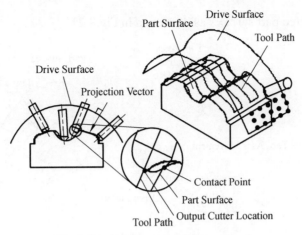

Fig.8-18 Surface area drive method

The drive surfaces can not be planar, but must be arranged in an orderly grid of rows and columns (see Fig.8-19). Adjacent surfaces must share a common edge and may not contain gaps that exceed the chaining tolerance defined under preferences. Trimmed surfaces can be used to define drive surfaces as long as the trimmed surface has four sides. Each side of the trimmed surface can be a single edge curve or comprised of multiple tangent edge curves that can be thought of as a single curve.

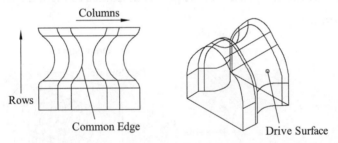

Fig.8-19 Even rectangular grid of rows and columns

The surface area drive method will not accept drive surfaces that are arranged in uneven rows columns or that have gaps exceed the Chaining Tolerance as illustrated in Fig.8-20.

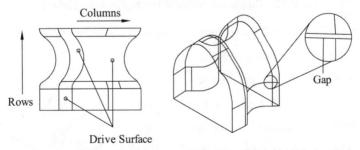

Fig.8-20 Uneven rows and columns

The surface area drive method provides maximum control of the tool axis. Variable tool axis options become available which allow you to define the tool axis relative to the drive surfaces. Additional tool axis control is sometimes necessary to prevent excessive tool undulations when machining very contoured part surfaces as illustrated in Fig.8-21.

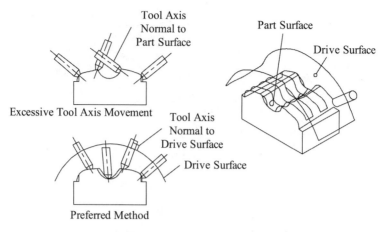

Fig.8-21　Tool axis movement using normal to drive

　　The surface area drive method also provides maximum control of the projection vector. An additional projection vector option, normal to drive, is available. This option enables you to evenly distribute drive points onto very convex part surfaces (Part Surfaces whose relative normals exceed 180 degrees). Unlike a boundary, drive surfaces can be designed to wrap around the part surfaces to project drive points evenly onto all sides of the part as illustrated in Fig.8-22.

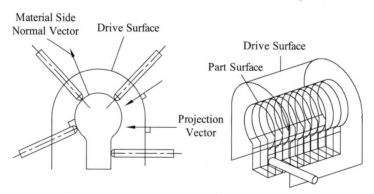

Fig.8-22　Projection vector is normal to drive surface

　　Once you have selected the drive surfaces, the system displays a default drive direction vector as illustrated in Fig.8-23. You may redefine the drive direction by selecting cut direction and then selecting one of eight displayed vectors which appear in pairs at each of the surface corners. The selected vector specifies the drive direction and the quadrant where the first cut will begin.

　　The options display in the Surface area drive method dialog box are shown in Tab.8-8.

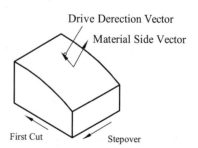

Fig.8-23　Material side and drive direction vectors

184

Tab.8-8 Surface area drive method options

Items	Description
Specify drive geometry	Specifies faces to define the drive geometry
Cut area	Defines how much of the total drive surface area can be utilized in the operation. These options are only available after you specify the drive geometry
Tool position	Specifies the tool position to determine how the software calculates contact points on the part surfaces.
Flip material	Reverses the material side direction vector of the drive surface
Display contact points	Displays a surface normal vector at each of the drive points generated for the operation
Cut step	Controls the distance between drive points along the cut direction
When gouging	Specifies how the software will respond when the tool gouges the drive surface during cutting moves

1. Tool position

· On: positions the tool tip at each drive point before projecting the tool path along the specified projection vector onto the part.

· Tanto: positions the tool tangent to the drive surface at each drive point before projecting the tool path along the specified projection vector onto the part.

On creates part surface contact points by first positioning the tip of the tool directly on the drive points and then projecting along the projection vector to the part surface, where the contact point is calculated. Fig.8-24 shows Tanto and On tool positions.

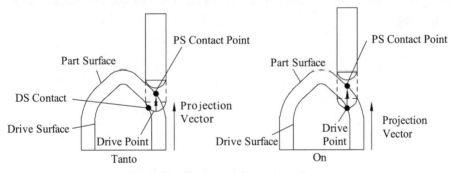

Fig.8-24 Tanto and On tool positions

Use tanto:

① For maximum part surface cleanup. You will get greater coverage on steep surfaces.

② When creating a tool path directly on the drive surface (with no part surface defined). Depending on the tool axis option, On may violate the drive surface as illustrated in Fig.8-25.

③ When the same surfaces have been defined as both the drive surfaces and part surfaces.

④ When the stepover option scallop is selected.

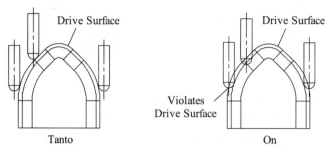

Fig.8-25　Tanto and on cutting drive surfaces

Since the cutting position on the tool may change and it is not necessarily represented by the tool path, the operation may appear to retrace, spike, or gouge. This condition may occur when you are machining a part where the radius of curvature of the part surface is smaller than the nose radius of the tool or when two surfaces meet at a concave corner. Changing your view of the tool path will reveal that the part surface is not being violated.

2. Cut step

Specify enough drive points along the drive cut path to capture the shape and features of the drive geometry; otherwise unexpected results may occur. When you machine directly on the drive surfaces and the tool axis is defined relative to the drive surfaces, specify more drive points to more accurately follow the contours of the drive surfaces.

・Number: Specifies the minimum number of drive points to create along the cut passes during tool path generation. Additional points are automatically created if necessary for the tool path to follow the part surface contours within the specified part surface Intol/Outtol values.

・Tolerances: Specifies Intol and Outtol values to define the maximum allowable normal distance between the drive surface and the line extending between two consecutive drive points, as shown in Fig.8-26.

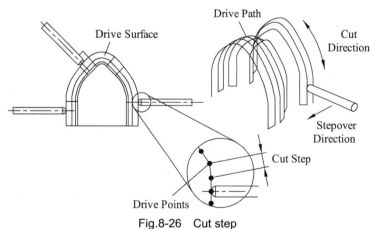

Fig.8-26　Cut step

3. When gouging

・None: Do not alter the tool path to avoid gouging and do not issue a warning message to the tool path or CLSF as shown in Fig.8-27.

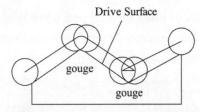

Fig.8-27 Generating Tool paths on Drive Surfaces using None

· Warning: Do not alter the tool path to avoid gouging but do issue a warning message to the tool path and CLSF.

· Skip: Alters the tool path by removing only the tool positions that cause gouging. The result is a straight tool movement from the last position before gouging to the first position which no longer gouges. The results of Skip not used and Skip used are shown in Fig.8-28.

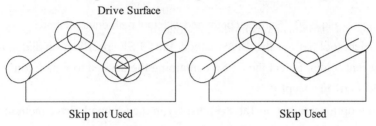

Fig.8-28 Skip used and not used

· Retract: Avoids gouging by using the engage and retract parameters defined in the Non-cutting moves dialog box.

8.3.7 Tool path drive method

Tool path drive method enables you to define drive points along the tool path of a Cutter Location Source File (CLSF) to create a similar surface contouring tool path in the current operation. Drive Points are generated along the existing Tool Path and then projected onto the selected part surfaces to create a new tool path that follows the surface contours. The direction of which the drive points are projected onto the part surfaces is determined by the projection vector.

In Fig.8-29, a tool path was created using planar milling, profile cut type. This tool path can be used by the tool path drive method operation to generate a new tool path that follows the contours of the part surface.

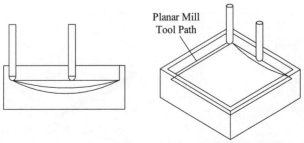

Fig.8-29 Planar milling, profile cut

187

Fig.8-30 illustrates the results of using the tool path drive method. The tool path created from the planar milling operation has been projected to the contoured part surface in the direction of the projection vector to create the surface contour tool path.

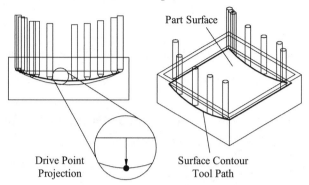

Fig.8-30　Surface contouring with tool path drive method

When you select tool path as the drive method, the CLSF specification dialog box displays the CLSF in your current folder. Select the CLSF containing the desired tool path (only one selection is allowed) and select OK to accept it.

The following options(shown in Tab.8-9) display in the tool path drive method dialog box.

Tab.8-9　Surface area drive method options

Items	Description
Tool paths	Lists the tool paths associated with the selected CLSF
Motion types by feedrate	Lists the feedrates associated with the various cutting and non-cutting moves in the selected tool path
Tool axis	Changes the tool axis specified earlier or specifies a tool axis available only in the surface area drive method
Projection vector	Determines how the drive points project to the part surface
Display drive path	Displays the current drive path used to create the tool path

8.3.8　Radial cut

The radial cut drive method enables you to generate drive paths perpendicular to and along a given boundary, using a specified stepover distance, bandwidth and cut type (see Fig.8-31). This drive method is useful in creating cleanup operations.

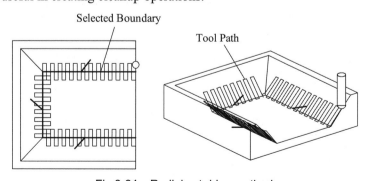

Fig.8-31　Radial cut drive method

Initially, the tool zigs or zig-zags along the boundary in the direction of the boundary indicators as illustrated in Fig.8-32. This can be changed by toggling from follow boundary to reverse boundary.

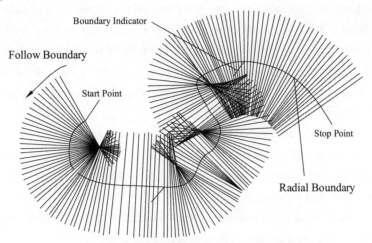

Fig.8-32　Direction is defined by boundary indicators

The following options (shown in Tab.8-10) display in the Radial cut drive method dialog box.

Tab.8-10　Surface area drive method options

Items	Description
Band on material/ opposite side	Defines the total width of the machined area measured in the plane of the boundary
Path direction	Determines the direction the tool travels along the boundary

Bandwidth defines the total width of the machined area measured in the plane of the boundary. The bandwidth is the sum of the band on material side and band on opposite side offset values.

The material side is the right-hand side of the boundary as you look in the direction of the boundary indicators illustrated in Fig.8-33. The opposite side is the left-hand side. The sum of the material side and opposite side cannot equal to zero.

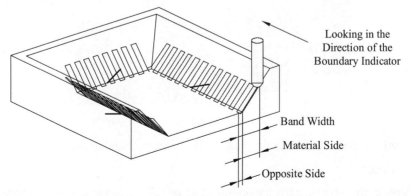

Fig.8-33　Material side and opposite side

189

8.3.9 Contour profile

The contour profile drive method in variable axis surface contouring machines canted walls with the side of the cutter. Variable axis profiling lets you automatically generate a tool path to machine the walls of a cavity or a region bounded by floor(s) and wall(s), with the sides of the cutter. After selecting the floor, the software can find all the walls that bound the floor. The tool axis is constantly adjusted to get a smooth path. At concave corners, the side of the tool is tangent to both adjacent walls. At convex corners, the software adds a radius and rolls the cutter around to keep the tool axis tangent to each corner wall.

Contour profile also allows you to machine walls that are not bounded by floors, such as the outside periphery of a part. There are two options to control the placement of the cutter against the wall when your part has no floors. Either use follow wall bottom to follow the periphery of the wall or use an auxiliary floor that behaves as a real floor.

Variable axis profiling is more automated than sequential milling, and is ideal for machining pockets with canted walls, such as those found in aerospace parts.

8.3.10 Flow cut

The Flow cut drive method is ideal for high speed machining, which removes excess material in corners prior to finishing or uncut material left behind by a previous, larger ball cutter. It generates tool paths one level at a time along the concave corners and valleys formed by part surfaces, and avoids embedding the tool when it moves from one side to the other.

The flow cut drive method calculates the direction and order of the cuts to optimize tool contact with the part and minimize non-cutting moves. It provides options to manually define the cut order and sequencing options(such as inside out) for operations with multiple offsets. These options help to generate a more constant cutting load and shorter non-cutting moves. It provides different cut patterns for steep areas and non-steep areas, as well as smooth turnarounds at the end of valleys as shown in Fig.8-34.

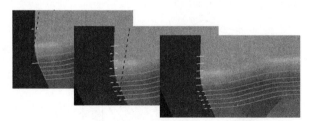

Fig.8-34 Smoothing improves turnarounds

The following options display in the flow cut drive method dialog box.

1. Drive geometry

・Max Concavity: Specifies the maximum angle of valleys to include in the current operation. For example, if you enter 120 in the max concavity box, the operation will machine the 110 and 70 degree valleys, but not the 160 degree valley. This is because it is shallow and flat. However, you are unable to machine all the material in the next two valleys, 110 and 70 degrees, respectively. More material is left in sharp corners and deep valleys like these.

When you return to machine the deep valleys missed by earlier passes, it is more efficient to machine just these deep valleys, and skip over the shallow valleys already machined in earlier passes. You can do this with maximum concavity by indicating which angles to ignore. In Fig.8-35, setting the maximum concavity angle to 120 degrees, for example, would machine the 110 and 70 degree valleys, but not the 160 degree valley.

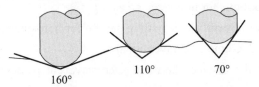

Fig.8-35 Max concavity

In Fig.8-36, the maximum concavity is set to 179 degrees. All angles less than or equal to 179 degrees are machined, in effect, cutting all valleys. In Fig.8-37, the maximum concavity is set to 160 degrees. All angles less than or equal to 160 degrees are machined.

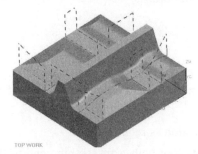 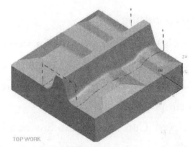

Fig.8-36   Concavity of 179 degrees   Fig.8-37   Concavity of 160 degrees

· Minimum cut length: Removes tool path cutting motions that are smaller than the specified length.

· Hookup distance: Unifies milling segments that are separated by less than the specified distance.

2. Drive Settings

· Flowcut type: Three flowcut types are available.

(1) Single pass: Generates one cutting pass of the tool along concave corners and valleys.

(2) Multiple passes: Generates multiple cutting passes from either side of the center flow cut, or from the inside outward. You specify a stepover distance to define the passes.

(3) Reference tool offsets: Generates multiple cutting passes from either side of the center flow cut, or from the inside outward.

This option is useful for cleanup machining after roughing out an area with a larger tool. You specify the reference tool diameter to define the total width of the area to machine, and a stepover distance to define the interior passes.

The smaller tool in the current operation removes the remaining material in the uncut regions where the larger reference tool could not fit.

3. Steep containment

Specifies the angle required to consider an area steep. The steep angle is measured between a horizontal plane and the tangent direction vector of the center flow cut. Enter a value between 0 and 90.

4. Non-steep cutting

Controls the cutting motions for non-steep areas.

·Non-steep cut pattern: Specifies the cutting pattern for non-steep areas.

·Cut direction: Mixed, Climb or Conventional.

·Stepover: specifies the distance between successive cut passes. The stepover distance is measured along the part surfaces.

·Number of stepovers per side: Specifies the number of passes to generate on each side of the center flow cut.

·Sequencing: Determines the order in which the zig, zig-zag, or zig-zag with lifts cut passes are executed. Available for multiple passes and reference tool offsets.

5. Steep cutting

Controls the cutting motions for steep areas.

·Steep cut pattern: Specifies the cutting pattern for steep areas.

·Steep cut direction: Available for the zig, crosscut zig, crosscut zig-zag, and crosscut zig-zag with lifts cut patterns.

(1) Mixed: Cuts in a high to low or a low to high direction as required.

(2) High to low: Machines the steep sections of the tool path from the high end to the low end.

(3) Low to high: Machines the steep sections of the tool path from the low end to the high end.

6. Reference tool

Available when flowcut type is set to reference tool offsets (see Fig.8-38).

·Reference tool diameter: Specifies the diameter of the tool that is used to determine the width of the finishing cut region.

·Overlap distance: Extends the width of the area to machine for the specified distance along the tangent surfaces of the remaining material.

The reference tool is typically the tool used previously to rough out the area. You must enter a diameter value that is larger than that of the current tool in use. The software calculates the bi-tangency contact points from the specified reference tool diameter and then uses these points to define the cut region for the finishing operation. The width of the area to machine is defined by the reference tool diameter.

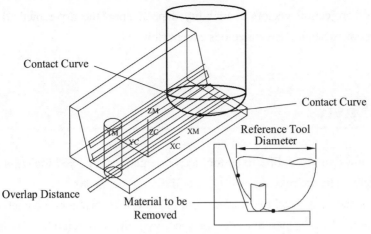

Fig.8-38 Reference tool

8.4 Projection Vector

Projection vector enables you to define how the drive points project to the part surface, and the side of the part surface the tool contacts. The surface area drive method provides one additional option, normal to drive that is not available for the other drive methods.

The drive points project along the projection vector to the part surface. Sometimes, as illustrated in Fig.8-39, drive point project in the opposite direction of the projection vector (but still along the vector axis) as they move from the drive surface to the part surface.

The direction of the projection vector determines the side of the part surface the tool contacts. The tool always positions to the part surface from the side the projection vector approaches. In Fig.8-39, drive point P_1 projects to the part surface in the opposite direction of the projection vector to create P_2.

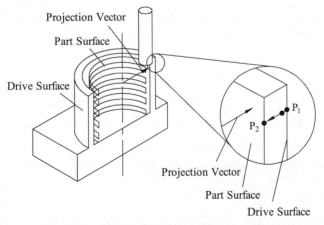

Fig.8-39 Drive point projects to part surface

The types of projection vectors availability depend upon the drive method. The projection vector option is common to all drive methods except flow cut.

8.5 Tool axis

Tool axis allows you to define fixed tool axis orientations. A fixed tool axis remains parallel to a specified vector. The are two options: +ZM or defining the Vector.

The tool axis can be defined by entering coordinate values, selecting geometry, specifying the axis relative or normal to the part surfaces, or specifying the axis relative or normal to the drive surfaces.

8.6 Create a fixed contouring operation

The following steps illustrate how to create a fixed contouring operation with curve/point drive method (see Fig.8-40).

(1) On the insert toolbar, click Create operation ., or choose insert → operation.

(2) In the Create operation dialog box, select mill contour from the type list and click Fixed contour in the operation subtype group.

(3) Select the required options in the location group.

(4) Click OK.

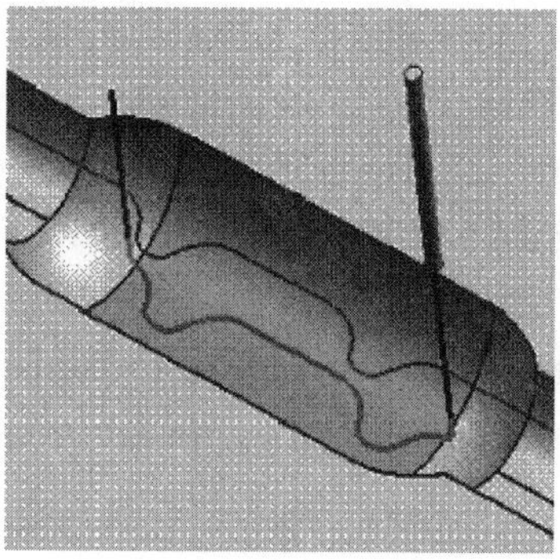

(5) (Optional) In the Fixed contouring dialog box, click Specify cut area and select the face or group of faces that require machining. If you do not specify a cut area, the entire part is considered the cut area.

Fig.8-40 Example of cut area with path

(6) In the drive method group, select curve/point from the method list.

In the Curve/Point drive method dialog box, select curve is active in the drive geometry group.

(7) Select the curves or face edges that need to be machined.

Alternately, click the Point constructor and select the points that define the area to be machined.

(8) Click OK.

(9) In the main Operation dialog box, click Generate .

8.7 Fixed contouring operation example

This example shows how to create fixed contouring operations using different drive methods. The machined model is shown in Fig.8-41:

Fig.8-41 Fixed contouring operation model

8.7.1 Create a fixed contour operation using the area milling drive method

(1) On the insert toolbar, click Create operation , or choose insert→operation.

(2) In the Create operation dialog box, from the type list, select mill_contour. In the operation subtype group, click Fixed contour .

(3) On the insert toolbar, click Create tool , or choose insert→tool.

(4) In the Create tool dialog box, from the type list, select mill_contour. In the tool subtype group, click ball mill . The software displays the Ball mill dialog box.

(5) In the Ball mill dialog box, set the tool parameters as shown in Fig.8-42.

(6) In the location group, set the options as shown in Tab.8-11.

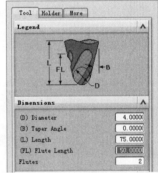

Fig 8-42 Ball mill tool dialog box

Tab.8-11 Operation options

Program	Program
Tool	Ball_mill_d4
Geometry	Workpiece
Method	Mill rough

(7) Click OK. The Fixed contour dialog box appears.

(8) In the geometry group, click Specify cut area . The Cut area dialog box is displayed.

(9) In the Selection options group, click Select all. The entire part will be machined.

(10) Click OK. The Fixed contour dialog box is displayed.

(11) In the drive method group, from the method list, select the area milling.

195

(12) Click Edit 🔧. The Area milling drive method dialog box appears (see Fig.8-43).

Fig.8-43 Area milling drive method dialog box

(13) In the drive settings group, from the cut pattern list, select follow periphery.
(14) From the cut direction list, select climb cut.
(15) From the stepover list, select scallop.
(16) In the scallop height box, type 3.
(17) From the stepover applied list, select on part.
(18) Click OK. The Fixed contour dialog box appears.
(19) Click Generate.
(20) Click OK to accept the tool path (see Fig.8-44) and close the dialog box.

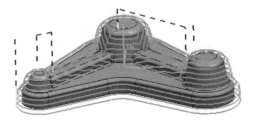

Fig.8-44 Fixed Contour Operation Path with Area Milling Drive Method

8.7.2 Create a flowcut multiple operation

(1) On the insert toolbar, click Create operation 📄, or choose insert→operation.
(2) In the Create operation dialog box, from the type list, select mill contour.
(3) In the operation subtype group, click Flowcut multiple 🔧.
(4) In the location group, set the options as shown in Tab.8-12.

Tab.8-12 Operation options

Program	Program
Tool	Ball_mill_d4
Geometry	Workpiece
Method	Mill finish

(5) Click OK. The Flowcut multiple dialog box appears as shown in Fig.8-45.

Fig.8-45 Flow multiple dialog box

(6) In the drive settings group, from the non-steep cut pattern list, select zig-zag.

(7) In the stepover box, type 1.

(8) In the number of stepovers per side box, type 5.

(9) In the actions group, click Generate .

(10) Click OK to accept the tool path (see Fig.8-46) and close the dialog box.

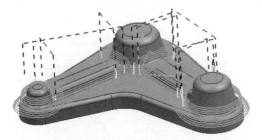

Fig.8-46 Flowcut multiple operation Path

8.7.3 Create a flowcut ref tool operation

This example uses a part that has already been roughed out with a 4.0 diameter ball mill cutter in a previous operation. A 2.0 diameter ball mill cutter is used to fit into areas where the 4.0 diameter ball mill cutter could not fit.

(1) On the insert toolbar, click Create operation , or choose insert→operation.

(2) In the Create operation dialog box, from the type list, select mill_contour.

(3) In the operation subtype group, click Flowcut ref tool .

(4) In the Create tool dialog box, from the type list, select mill contour. In the tool subtype group, click Ball mill . The software displays the Ball mill dialog box.

(5) In the Ball mill dialog box, set the tool parameters as shown in Fig.8-47.

197

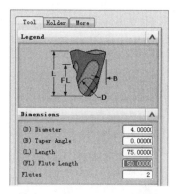

Fig.8-47 Ball mill tool dialog box

(6) In the location group, set the options as required (see Tab.8-13).

Tab.8-13 Operation options

Program	Program
Tool	Ball_mill_d2
Geometry	Workpiece
Method	Mill finish

(7) Click OK. The Flowcut ref tool dialog box appears as shown in Fig.8-48.

Fig.8-48 Flowcut ref tool dialog box

(8) In the steep group, from the containment list, select non-steep.

(9) In the drive settings group, from the cut patterns list, select zig-zag.

(10) In the reference tool group, in the reference tool diameter box, type 4.0.

The 4.0 diameter ball mill cutter used in the roughing operation is the tool being referenced.

(11) In the reference tool group, in the overlap distance box, type 1.0.

(12) In the actions group, click Generate.

(13) In the actions group, click Verify. The Tool path visualization dialog box appears.

(14) Click the 2D dynamic tab.

(15) Click Play.

(16) Click OK. The Flowcut ref tool dialog box appears.

(17) In the actions group, click OK to accept the tool path (see Fig.8-49) and close the dialog box.

Fig.8-49　Flowcut Ref Tool Operation Path

 Notes

[1] Selecting the appropriate drive method should be determined by the shape and complexity of the surface you wish to machine, and the tool axis and projection vector requirements. The selected drive method determines the type of drive geometry you can select, as well as the available projection vectors, tool axis, and cut types.

选择合适的驱动方式应取决于待加工表面的形状和复杂程度、刀轴及投影矢量的需求。选中的驱动方式决定了你能够选择的驱动几何体，以及投影矢量、刀轴和切削类型。

[2] Unlike other drive methods which require an abrupt change in direction to stepover to the next cutting pass, Spiral drive method stepovers are a smooth, constant transition outward. Because this drive method maintains a constant cutting speed and smooth motion, it is useful for high speed machining applications.

和其他在步进到下一个切削刀路时需要在方向上发生陡变的切削方式不同，螺旋线驱动方式的步进是一个平缓的恒定的向外过渡过程。由于该种驱动方式保持着一个恒定的切削速度和平缓运动，因此在高速加工应用中很有用。

[3] The area milling drive method enables you to define a fixed-axis surface contouring operation by specifying a cut area and, if desired, adding steep containment and trim boundary constraints. It is similar to the boundary drive method, but requires no drive geometry and uses a robust and automated computation of collision-free containment.

区域铣削驱动方式运行用户通过限定切削区域定义一个固定轴表面轮廓操作，并在有需要的情况下，增加陡角包容和修剪边界限制。它和边界驱动方式类似，但不需要驱动几何体，采用无冲突包容的鲁棒自动运算。

[4] The flow cut drive method is ideal for high speed machining, which removes excess material in corners prior to finishing or uncut material left behind by a previous, larger ball cutter.

对高速加工来讲，清根驱动是一种理想的方式，该种方式在拐角处去除大量材料，这些材料是由之前大的球铣刀遗留下的未切削材料。

New Words in Chapter 8

projection	[prə'dʒekʃən]	n.	投影，投射
groove	[gruːv]	n.	凹槽，沟槽
undulation	[ˌʌndʒə'leʃən]	n.	波动；起伏
normals			法线，法向
quadrant	['kwɒdrənt]	n.	象限，象限角
spike	[spaɪk]	n.	尖峰
bandwidth	['bændwɪdθ]	n.	带宽
concavity	[kɒn'kævəti]	n.	凹度，凹角
hookup	['hʊkˌʌp]	n.	连接

Chapter 9
Drilling

Objectives:
- ✓ To understand the drilling operation.
- ✓ To understand the drilling cycles and cycle parameters of drilling operation.
- ✓ To be able to create a drilling operation.

9.1 Drilling operations overview

9.1.1 Drilling operations introduction

Drilling operations (see Fig.9-1) are used to machine simple holes, which use a point to point processor to generate tool paths. For more complex holes such as counterbore holes or undercut holes, you can consider using milling operations.

Drilling operations are associated with drill geometry parent groups for which only the hole position, top surface and bottom surface need to be specified. For multiple holes of the same diameter in a part, you can create the drilling operations rapidly by specifying different drilling cycles and cycle parameters rather than specifying each hole individually, which will make the tool changing and repositioning unnecessary, so as to save considerable machining time and improve machining efficiency.

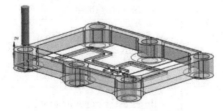

Fig.9-1 Drilling operation

Several drill operation subtypes are available, in which a tool is positioned to geometry, fed into the part, and retracted. You can define parameters such as:
- Depth.
- Minimum clearance distance.

- Dwell.
- Break chip.

9.1.2 Drill Operation subtypes

Fourteen drill operation subtypes are available, as shown in Tab.9-1, which all use the *drill* point-to-point processor for removing material when manufacturing holes, with the exception of mill control, mill user, and thread milling.

The machining holes include but not limited to through hole, blind hole, counterbore and countersink.

Tab.9-1 Drill operation subtypes

Icon	Subtype	Description
	Spot facing	Allows the tool to pause at the depth of cut by a specified number of seconds or revolutions
	Spot drilling	Allows the tool to pause at the tool tip or shoulder depth of the tool by a specified number of seconds or revolutions
	Drilling	Allows users to do basic point-to-point drilling
	Peck drilling	Allows users to create a series of drilling motions into and out of a hole at progressive intermediate increments. The tool retracts out of the hole after each peck to clear the chips
	Breakchip drilling	Allows the drilling tool to retract slightly after feeding to an incremental depth to break the chip
	Boring	Allows the boring tool to feed continuously in and out of the part
	Reaming	Allows the reamer to feed continuously in and out of the part
	Counterboring	Allows the tool to pause at the depth of cut by a specified number of seconds or revolutions. Use this operation subtype when you need a dwell in your counterboring cycle
	Countersinking	Allows the tool to pause at the depth of cut by a specified number of seconds or revolutions. Use this operation subtype when need a dwell in countersinking cycle
	Tapping	Creates a tap cycle that feeds into the hole, reverses the spindle, and feeds out of the hole
	Thread milling	Mills threaded holes with helical cuts
	Mill control	Allows users apply machine control events such as coolant on/coolant off or spindle clockwise/counterclockwise
	Mill user	Allows users generate tool path using custom nx open program

9.1.3 Drill tool

When you set type to drilling tool, select the subtype of the tool you want to create. Tab.9-2 is a list of drill tools and the key dimensions that you can set for each tool.

These tools are intended for use with the point-to-point module. The point-to-point module only recognizes a drilling tool. Therefore, the spot drill, boring bar, reamer, counterbore, countersink, and tap are all seen by the system as the standard drill. These subtypes are available to help you separate and organize your tools.

Tab.9-2 Drill tool subtypes

Tools Type	Icon	Drafts	Parameter Description
Spotfacing tool			(D) Diameter (R1) Lower radius (L) Length (FL) Flute length
Spotdrilling tool			(D) Diameter (R1) Lower radius (L) Length (FL) Flute length
Drilling tool			(D) Diameter (R1) Lower radius (L) Length (FL) Flute length
Boring bar			(D) Diameter (CR) Corner radius (L) Length (FL) Flute length
Reamer			(D) Diameter (L) Length (PA) Point angle (FL) Flute length
Counterboring tool			(D) Diameter (L) Length (A) Tip angle (FL) Flute length
Countersinking tool			(D) Diameter (L) Length (A) Tip angle (FL) Flute length
Tap			(D) Diameter (L) Length (PA) Point angle (FL) Flute length
Thread mill			(SD) Shank diameter (D) Diameter (L) Length (FL) Flute length (P) Pitch

9.2 Drill geometry

9.2.1 Drill geometry parent groups

Use drill geometry parent groups to manage the drill geometry. If drilling operations are associated to drill geometry parent groups, same geometry rather than a new one can be used for each operation.

Drill geometry parent groups from which drilling operations can inherit information, are shown in Tab.9-3.

Tab.9-3 Drill geometry parent groups

Icon	Subtype	Description
	MCS	Defines the origin for subsequent tool path data based on the MCS
	Workpiece	Can inherit the MCS from the MCS parent group. You can assign part and blank material. This parent group is also used in tool path verification
	Drill geom	Defines hole geometry used in drilling operations. It can inherit the workpiece from the workpiece parent group

9.2.2 Drill geometry subtypes

Three drill geometry subtypes are available, as shown in Tab.9-4. You can use any of them to specify holes for drilling operations.

Tab.9-4 Drill geometry subtypes

Icon	Subtype	Description
	Specify holes	Select the hole geometry and specify parameters
	Specify top surface	Define a face or plane to which all points are projected
	Specify bottom surface	Define the hole depth in the model

9.2.3 Create drill geometry parent groups

Drill geometry parent groups are used to manage the drill geometry. An example is given below to show how to create a drill geom parent group that defines hole geometry that can be inherited in multiple drilling operations.

(1) On the insert toolbar, click Create geometry , or choose insert→geometry.

(2) In the Create geometry dialog box, in the type group, from the list, select drill.

(3) In the geometry subtype group, click Drill geom .

(4) In the location group, from the geometry list, select a geometry parent group. This example uses workpiece because the part geometry was defined in the workpiece.

(5) Click OK. The Drill geom dialog box appears.

(6) In the geometry group, click Specify holes 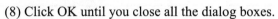. The Point dialog box appears.

(7) Click Select. In the graphics window, select the edge of the holes as shown in Fig.9-2.

(8) Click OK until you close all the dialog boxes.

The drill geometry parent group is created, and is displayed in the geometry view in the Operation Navigator.

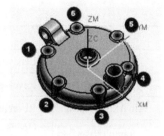

Fig.9-2　Create drill geometry parents

9.2.4　Specify holes

Specify holes is applied to select the hole geometry and specify parameters., as shown in Fig.9-3.

When you click Specify holes , the Point dialog box is displayed, as shown in Fig.9-4. The Point dialog box contains several options as shown in Tab.9-5, which allow users to select and manipulate points for generating a tool path.

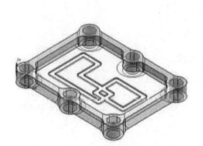

Fig.9-3　Specify holes　　　　Fig.9-4　Point dialog to specify hole

Tab.9-5　Point dialog box options

Option	Description
Select	Opens a selection dialog box to let you select geometry that represents the top of the hole
Append	Adds points that you select to previously selected drill geometry
Omit	Ignores points that you select on previously defined drill geometry
Optimize	Opens a dialog box where you can set options to optimize tool travel to reduce the length of the tool path
Display Points	Displays the new order of points

205

Contiinue

Option	Description
Avoid	Lets you specify tool clearance over fixtures or obstacles within the part
Reverse	Reverses the order of previously selected goto points
Arc Axis Control	Reverses the tool axis orienatation for selected arcs and holes in sheet bodies
Rapto Offset	Opens a dialog box where you can specify a RAPTO value to each selected point, arc, or hole
Planning Complete	Closes the Point dialog box, and returns the Drill geom dialog box

1. Select

Only the select option is used when specifying geometry. The remaining options let you edit already selected geometry. When the Select hole dialog box (shown as Fig.9-5) is open, you can select the geometry in the graphics window, by name, or using any of the available options, as shown in Tab.9-6.

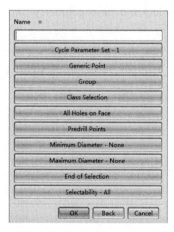

Fig.9-5 Select hole dialog

You can use the following different types of geometry to represent a hole in a point-to-point operation:

· A cylindrical hole or conic hole in either a sheet body or a solid body.
· A point.
· An arc or full circle.

Tab.9-6 Select hole dialog box

Option	Description
Cycle parameter set	Defines which cycle parameter is set to associate with the next set of points
Generic point	Lets you specify drill points using the Point constructor dialog box
Group	Lets you reference a previously defined group of points and arcs. You must enter the name of the group
Class selection	Opens the Class selection dialog box. You can use the filtering options to select the points and arcs you want

Continue

Option	Description
All Hole on Face	Selects all the holes on the specified face
Predrill points	Lets you reference pre-drills holes that were specified earlier in a planar milling and cavity milling operation
Minimum diameter	Lets you specify a diameter value to restrict the selection of arcs to those with a diameter less than or equal to the specified value
Maximum diameter	Lets you specify a diameter value to restrict the selection of arcs to those with a diameter greater than or equal to the specified amount
End of selection	Closes the dialog box and returns to the Point dialog box
Selectaed bility	Lets you filter for the type of geometry that you want to select: · Points only · Arcs only · Holes only · Points and arcs · All

2. Append

The append option makes the Point dialog box to be displayed, which allows users to add points to previously selected drill geometry.

3. Omit

The omit option ignores points that you select on previously defined drill geometry.

4. Optimize

The optimize option allows you to optimize tool travel to reduce the length of the tool path (see Tab.9-7). When you click it, the Optimize dialog box opens, as shown in Fig.9-6.

Fig.9-6 Optimize dialog

Tab.9-7 Optimize dialog box option

Option	Description
Shortest path	Arranges the points in the order required to minimize total machining time
Horizontal bands	Let you restrict the tool path with horizontal bands, which are used for other machining constraints such as clamp locations, machine travel limits, table size, and so on
Vertical bands	Let you restrict the tool path with vertical bands these bands
Repaint points	Repaints all the points after each optimization if set to yes

207

The shortest path option is the most commonly used, which arranges the points in the order required to minimize total machining time. The Optimization dialog box is shown in Fig.9-7.

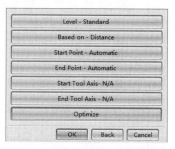

Fig.9-7   The shortest path optimization dialog

· Level *standard* refers to the process of analysis that you want to use in determining the shortest tool path and *advanced* increases machining time efficiency at a maximum.

· Based on is the only option for a fixed axis tool path, while variable axis tool paths can take the tool axis into account when determining machining efficiency.

· Start point controls the start point of the tool path.

· End point controls the end point of the tool path.

· Start tool axis controls the tool axis at the beginning of the cutting motion, which is only available for variable axis tool paths.

· End tool axis controls the tool axis at the end of the cutting motion, which is only available for variable axis tool paths.

· Optimize initiates the optimization process.

5. Display Points

The display points option allows you to display the new order of points after using the include, omit, avoid or optimize options. Points after using omitting option are shown in Fig.9-8

(a) Original point order   (b) New point order after point omitted

Fig.9-8   Display points after using omitting option

6. Avoid

The avoid option (see Fig.9-10) lets you specify tool clearance over fixtures or obstacles within the part, as shown in Fig.9-9. You must define start point, end point and avoid distance. Avoid distance represents the distance between the part surface and the tool tip.

Avoid motion is generated by the point-to-point module so that the tool path contains goto statements to move the tool up and over obstacles. It is associated with pairs of adjacent points. The

NX generates motions to avoid obstacles between each pair of adjacent points.

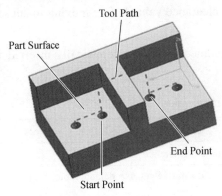

Fig.9-9　Set avoid parameter

Fig.9-10　Avoid clearance dialog

7. Reverse

The reverse option reverses the order of previously selected goto points, which is used to perform back-to-back operations, such as drill and tap, on the same set of points. The process allows you to start the second operation where the first operation ends.

8. Arc axis control

The arc axis control is used to reverses the tool axis orientation for selected arcs and holes in sheet bodies, as shown in Fig.9-11.

· Display apply to display the tool axis orientation of arcs and holes by single or all, as shown in Fig.9-12.

· Reverse reverses the tool axis orientation for selected arcs and holes.

Fig.9-11　Arc axis control dialog

Fig.9-12　Display dialog

9. Rapto offset

The rapto offset is the location where the feedrate changes from rapid feed rate to a cut feed rate.

You can specify a rapto value in the dialog box shown in Fig.9-13. The options of rapto value are listed in Tab.9-8.

Fig.9-13　Rapto offset dialog

The rapto value can be positive or negative. If you specify a negative rapto:

· The tool retracts out of holes to the specified clearance value before moving to subsequent hole locations.

· If no RTRCTO is specified, a RTRCTO=Clearance value is automatically inserted for all drill cycles except the manual bore cycle.

Tab.9-8 Specify a rapto value

Option	Description
Default rapto	Sets the offset to the positive minimum clearance
Rapto offset	Displays the current offset. If you do not select use default rapto, you can enter another value for the selected hole
Select all	Lets you select either all holes, or only holes that are of the same type and size or on the same face
Deselect all	Lets you deselect all previously selected holes
Display holes	Shows the tool path sequence number of each hole
List raptos	Outputs each hole rapto information to the listing device
Filters	Lets you filter what is displayed and listed. All and non-default apply to both Display Holes and List raptos. Non-Default displays and lists only those holes not using the default rapto

9.2.5 Specify top surface

Top surface (see Fig.9-14) is used to define a face or plane to which all points are projected. It can be an existing face of the part, or an ordinary plane. If top surface is not specified, the NX defaults to the location of selected point. Your drilling feedrate starts at the minimum clearance.

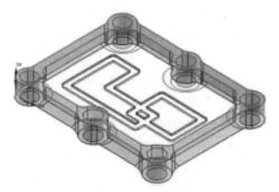

Fig.9-14 Top surface

9.2.6 Specify bottom surface

Bottom surface (see Fig.9-15) is used to define the hole depth in the model. You must select a face, a plane, or an arc to specify the bottom surface to drill to.

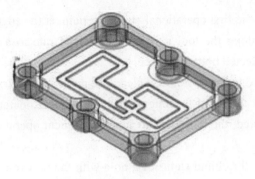

Fig.9-15 Bottom surface

9.3 Drilling cycles

9.3.1 Drilling cycles overview

A drilling cycle describes the machine tool movements necessary to perform point-to-point machining functions, such as drilling, tapping, or boring. Through postprocessing, cycle statements are normally output as canned cycle codes for all the tool motions in a cycle operation to be executed by the machine tool.

In machines without canned cycle codes, goto command statements are used to define each of the tool motions and machine functions which simulate the specified cycle.

The following are some characteristics of cycle types:

· No cycle, peck drill, and break chip do not output cycle commands in the tool path. The motion is simulated with goto points.

· Standard cycle options output a cycle command at each of the specified CL (Cutter Location) points.

· Standard drill, standard drill csink, standard drill deep and standard drill break chip output canned cycles and are the equivalent of no cycle, peck drill and break chip which output simulated motion.

9.3.2 Cycle list

Fourteen options appear in the cycle type group in Drilling dialog boxes, as shown in Fig.9-16.

1. No cycle

It cancels any active cycle. If no cycle is active, when you generate a tool path, the tool motions are generated in the following sequence:

Fig.9-16 Cycle list

(1) The tool moves to the first operational clearance point at the engage feedrate.

(2) The tool moves along the tool axis at the cut feed rate to a point that allows the tool shoulder to go beyond a selected bottom surface.

(3) The tool is retracted to the operational clearance point at the retract feedrate.

(4) The tool moves at the rapid feedrate to each subsequent operational clearance point. If a bottom surface is not selected, the tool feeds to each subsequent operational clearance point at the cut feedrate.

You cannot define the thru/blind status of a hole with the no cycle option. However, you can use the peck drill or breakchip cycle options, set the depth to thru bottom surface, bottom surface, or model depth, and set the increment parameter to none to get the desired results.

2. Peck drill

It generates a simulated peck drilling cycle at each selected CL-point. A peck drilling cycle consists of a series of drilling motions into and out of a hole at progressive intermediate increments. The software uses goto/ command statements to describe and generate the required tool motions in the following order:

(1) The tool moves at the Cycle feed rate to the first intermediate increment.

(2) The tool moves from the hole to the operational clearance point at the retract feedrate.

(3) The tool moves at the engage feedrate to a clearance point above the previous depth.

(4) The tool moves at the cycle feedrate to the next intermediate depth which you set using the Increment option. The increment can be none, constant, or variable. This series of motions continues until the specified hole depth is reached, at which point the software removes the tool from the hole to the operational clearance point at the retract feedrate.

(5) The tool positions itself at the subsequent drilling location at the rapid feedrate, and begins the next drilling cycle.

3. Break Chip

It generates a simulated break chip drilling cycle at each selected CL-point. The break chip drilling cycle is identical to a peck drilling cycle with the following exception: after each drilling increment, instead of generating a full retract motion out of the hole and a return motion back to a distance above the previous depth, the software generates a retract motion to a point that is the distance above the current depth.

The software generates the following specific tool motions:

(1) The tool moves along the tool axis at the Cycle Feed Rate to the first intermediate increment.

(2) The tool moves back from the current position at the Retract Feed Rate to a clearance point above the current depth, defined by the Distance prompt. (see the caution below.)

(3) The tool moves at the Cycle Feed Rate to the next intermediate depth.

(4) This series of motions continues until the specified hole depth is reached, at which point the software removes the tool from the hole to the operational clearance point at the Retract Feed

Rate.

(5) The tool positions itself to subsequent operational clearance points at the Rapid Feed Rate.

4. Standard text

It activates a cycle/ statement with positioning motions according to the specified APT command statement. This statement can contain only minor words and parameters. Enter the words or parameters exactly as you want them to be output following the slash.

5. Standard drill

It causes the software to generate a standard countersink cycle at each selected CL-point. The software generates a cycle statement of the form: CYCLE/DRILL, CSINK, CSKDIA, d, TLANGL, a, parameters where

(1) The words in uppercase letters are standard APT vocabulary words.

(2) d represents the countersink diameter.

(3) a represents the tool point angle specified in the tool data definition.

(4) Parameters represent the appropriate supporting cycle parameters.

The resultant motion is dependent upon the target machine tool and postprocessor. However, a typical CYCLE/DRILL, CSINK sequence consists of a feed to a depth, calculated from the countersink diameter and tool point angle, and a retract along the tool axis at the rapid feedrate.

6. Standard drill, csink

See standard drill.

7. Standard drill, deep

It activates a standard deep drilling cycle at each selected CL-point. The software generates a cycle statement of the form: CYCLE/DRILL, DEEP, parameters where

(1) The upper case words are standard APT vocabulary words.

(2) Parameters represent appropriate supporting cycle parameters.

Resultant motions are dependent upon the target machine tool and postprocessor. However, a typical deep drilling sequence consists of a feed to depth in a series of increments, with a retract out of the hole after reaching each new incremental depth. The retract is at the rapid feedrate.

8. Standard drill, brkchp

It activates a standard break chip drilling cycle at each selected CL-point. The software generates a cycle statement of the form: CYCLE/DRILL, BRKCHP, parameters where

(1) The words in upper case letters are standard APT vocabulary words.

(2) Parameters represent the appropriate supporting cycle parameters.

Resultant motions are dependent upon the target machine tool and postprocessor. However, a typical break chip drilling sequence consists of a feed to depth in a series of increments with a retract to a safe clearance distance between each increment, and a retract out of the hole after the final depth has been reached. The clearance distance is defined by your controller or machine tool. The retract is at the rapid feedrate.

9. Standard tap

It activates a standard tap cycle at each selected CL-point. The software generates a cycle statement of the form: CYCLE/TAP, parameters where

(1) The words in upper case letters are standard APT vocabulary words.

(2) Parameters represent the appropriate supporting cycle parameters.

Resultant motions are dependent upon the target machine tool and controller. However, a typical tap sequence consists of a feed to depth, spindle reverse, and feed back out of the hole.

10. Standard bore

It activates a standard bore cycle at each selected CL-point. The software generates a cycle statement of the form: CYCLE/BORE, parameters where

(1) The words in upper case letters are standard APT vocabulary words.

(2) Parameters are the appropriate supporting cycle parameters.

Resultant motions are dependent upon the target machine tool and postprocessor. However, a typical bore sequence consists of a feed to depth and feed back out of the hole.

11. Standard bore, drag

It activates a standard bore cycle with a non-rotating spindle retract at each selected CL-point. The software generates a cycle statement of the form: CYCLE/BORE, DRAG, parameters where

(1) The words in upper case letters are standard APT vocabulary words.

(2) Parameters represents the appropriate supporting cycle parameters.

Resultant motions are dependent upon the target machine tool and postprocessor. However, a typical bore drag sequence consists of feed to depth, spindle stop, and retract out of the hole. The retract is at the rapid feedrate. When you select this option, the software prompts you for the number of cycle parameter sets you intend to use.

12. Standard bore, nodrag

It activates a standard bore cycle with spindle stop and orient at each selected CL-point. The software generates a cycle statement of the form: CYCLE/BORE, NODRAG, q, parameters where

(1) The words in upper case letters are standard APT vocabulary words.

(2) q represents the spindle orient value.

(3) Parameters represents the appropriate supporting cycle parameters.

Resultant motions are dependent upon the target machine tool and postprocessor. However, a typical bore nodrag sequence consists of:

(1) A feed to depth.

(2) Spindle stop and orient.

(3) Offset motion perpendicular to the tool axis and in the direction of the spindle orient.

(4) Rapid retract out of the hole.

13. Standard bore, back

It activates a standard back boring cycle at each selected CL-point. The software generates a cycle statement of the form: CYCLE/BORE, BACK, q, parameters where

(1) The words in upper case letters are standard APT vocabulary words.

(2) q represents the spindle orient value.

(3) Parameters represents the appropriate supporting cycle parameters.

Resultant motions are dependent upon the target machine tool and postprocessor. However, a typical back boring sequence consists of:

(1) A spindle stop and orient.

(2) Offset motion perpendicular to the tool axis and in the direction of the spindle orient.

(3) Dead spindle feed into the hole.

(4) Offset motion back to the center of the hole.

(5) Spindle start.

(6) Feed back out of the hole.

14. Standard bore, manual

It activates a standard bore cycle with manual spindle retract at each selected CL-point. The software generates a cycle statement of the form: CYCLE/BORE, MANUAL, parameters where

(1) The words in upper case letters are standard APT vocabulary words.

(2) Parameters represents the appropriate supporting cycle parameters.

Resultant motions are dependent upon the target machine tool and postprocessor. However, a typical bore manual sequence consists of a feed to depth, spindle stop, and program stop to allow the operator to manually retract the spindle out of the hole.

9.3.3 Minimum clearance

The Minimum clearance distance (see Fig.9-17) determines how the tool is positioned before entering the material.

· If a clearance plane is not specified, the tool positions to the next hole at the rapid feedrate directly to the specified minimum clearance distance above the part surface.

· If a clearance plane is specified, the tool moves at the rapid feed rate from the clearance plane to the specified minimum clearance.

· After the operations are completed, the tool moves back to the clearance plane.

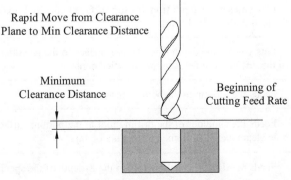

Fig.9-17　Minimum clearance

9.4 Cycle Parameters

Cycle parameters define exact tool motions and conditions such as feedrates, dwell times, and cutting increments. The dialog varies slightly as the different cycle type. A typical Cycle parameters dialog is shown in Fig.9-18. If you want to vary any of the cycle parameter values, a cycle parameter set for each hole or group of holes can be created. You can have up to five parameter sets per cycle.

1. Depth

Opening the Cycle depth dialog box (as shown in Fig.9-19) where you can specify the depth of cut using one several options. This value represents the total depth of a hole from the part surface to the tool tip. This option is not available for standard drill csink.

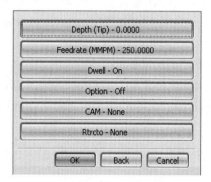

Fig.9-18 Cycle parameters dialog

Fig.9-19 Cycle depths dialog

The following options (see Tab.9-9) display in the Cycle depth dialog box, shown as Fig.9-20.

Tab.9-9 Cycle depth dialog box

Option	Description
Model depth	Computes the depth for each hole in a solid model. The tool axis must be the same as the hole axis
Tool tip depth	Lets you set the depth from the top surface to the tool tip. The tip of the drilling tool feeds to the specified depth
Tool shoulder depth	Lets you set the depth from the top surface to the shoulder of the tool. The shoulder of the drilling tool feeds to the specified depth
To bottom surface	Feeds the tool tip to the bottom surface
Thru bottom surface	Feeds the shoulder of the drilling tool to the bottom surface. You can specify a thru hole clearance value in the Drilling dialog box
To selected point	Feeds the tip of the drilling tool to the Z-depth of the specified point

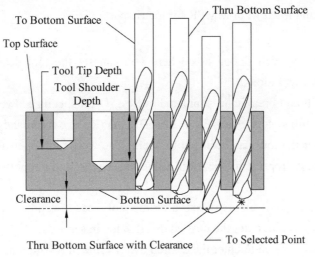

Fig.9-20 Cycle depth

Depth offsets (see Fig.9-21) options are available in drilling dialog boxes when you need to set the following for the operation:

① Blind hole stock. You can specify the distance above the bottom of a blind hole at which the tool is to stop drilling.

② Thru hole clearance. You can specify the distance to which the tool travels beyond the bottom surface of the through hole.

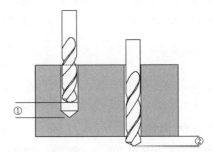

Fig.9-21 Depth offsets

The depth offset options are used in conjunction with the depth option in the Cycle parameters dialog box as follows:

· If the depth option is set to model depth, the depth offset applies only to solid holes. It does not apply to points, arcs, or holes in sheet bodies.

· If the depth option is set to to bottom surface, the blind hole stock is applied to all selected objects. A bottom surface must be active.

· If the depth option is set to thru bottom surface, the thru hole clearance is applied to all selected objects. A bottom surface must be active.

2. Feedrate

Opening the Cycle feedrate dialog box where you can specify the transit speed of tool while cutting. The current feedrate units and value are displayed. You can set the feedrate to IPM or IPR for non-metric parts and MMPM or MMPR for metric parts.

3. Dwell

Opening the Cycle dwell dialog box where you can set the tool delay at depth of cut by number of seconds or revolutions.

For the peck drill and break chip simulated cycles, if you specify a dwell value in seconds or revolutions, the software generates a DELAY/t or DELAY/REV, r commands statement to activate the desired dwell after the tool has been fed to depth. The words in uppercase letters are standard APT vocabulary words; t represents the dwell value in seconds, and r represents the dwell value in revolutions.

· Off specifies that no dwell is to occur after the tool feeds to depth.

· Seconds allows you to enter the desired dwell value in seconds.

· Revolutions allows you to enter the desired dwell value in revolutions of the spindle.

· For the standard cycles, if you specify a dwell value in seconds or revolutions, the software includes a DWELL, t or DWELL, REV, r parameter in the cycle/ statement to activate the specified cycle with the specified dwell value. You have an additional option on for standard cycles. If you use this option the word dwell is included in the cycle statement. This causes the tool to dwell in place for the specified time period at the depth of cut.

4. Option

Turning on or off machining characteristics that are unique to a particular machine. This option is postprocessor dependent. If you turn this option on, the software includes option in the cycle/ statement.

5. CAM

Specifying a preset CAM stop position for tool depth for machine tools with no programmable Z-axis.

· This value must be a positive integer or zero. If you enter a positive integer value, v, the software includes a CAM,v parameter string in the cycle/ statement. The postprocessor, in turn, generates code to feed the tool to CAM stop position v.

· If you enter zero, no CAM parameter is included in the cycle/ statement, and the CAM status appears as none on the cycle parameters menu.

· If you enter a negative value, the software displays the message invalid CAM value.

· Specify a CAM value only if the CAM parameter is valid for the target postprocessor.

Even though the CAM parameter controls tool depth for machine tools for which the Z-axis is not programmable, Siemens PLM Software recommends that you define the depth parameter as well so that feed-to-depth motion can be displayed.

6. Csink diameter

Lets you set the diameter of a countersink hole as shown in Fig.9-22. The software includes the countersink diameter you specify as a CSKDIA, d parameter on the cycle/drill csink statement.

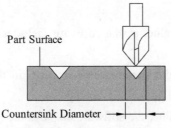

Fig.9-22　Csink Diameter

7. Entrance diameter

Lets you specify the outside diameter of an existing hole which is to be enlarged by a countersink operation.

Entrance diameter represents the diameter of an existing hole before countersinking. This option is available only for the standard drill, csink cycle.

The entrance diameter d specifies the diameter of the existing hole which is to be countersunk. A non zero value causes the software to include a holdia, d parameter in the cycle statement. The postprocessor typically uses this parameter to calculate a rapid positioning point inside the hole. The tool moves to a clearance point above the centerline of the hole before moving into the hole.

8. Increment

Lets you specify the dimensional value of one of a series of regular consecutive cuts to progressive depths used in peck and break chip drilling operations.

· None feeds the tool to depth in one motion.

· Constant lets you enter a positive constant increment value. The software generates motion to feed the tool in a series of constant increments until the tool reaches the final depth. If the increment is not equally divisible into the specified depth, the software uses the increment distance as many times as possible without drilling beyond the specified depth. The software then calculates the remaining distance and causes the tool to drill to the specified depth.

· Variable lets you define up to seven unique increment values. Except for the seventh increment, specify a non-zero value for the repeat factor. The software generates the specified number of repeats and then goes to the next increment.

If you enter a value of zero for an increment, the software ignores that increment and all subsequent increments.

For the last non-zero increment, the software generates the required number of repeats to feed the tool to the specified depth. In no case does the software cause the tool to feed below the specified depth.

9. Rtrcto

Lets you specify a retract distance. The distance is measured along the tool axis, from the part surface to a point to which the tool retracts after feeding to depth. This option is available for all standard cycles except standard bore, manual.

Distance lets you specify a distance for the retraction of the tool. You can enter a positive value, zero, or accept the default.

(1) If you enter a positive value, *d*, the software includes a rtrcto,d parameter string in the cycle/ statement.

(2) If you enter zero, no rtrcto parameter is included in the cycle/ statement and the rtrcto status appears as none in the Cycle parameters dialog box.

(3) If you enter a negative value, the software displays the message invalid rtrcto value. Click OK in the message box. Enter a non-zero rtrcto value only if you want the tool to be retracted to a point above the operational clearance point specified by the minimum clearance value.

Auto retracts the tool to the pre-cycle position for the current parameter set and successive parameter sets that use the auto option. The tool retracts along its tool axis to its last position before the current cycle.

This option automatically sets the rtrcto value for all parameter set points.

Set to none omits the use of a retract distance.

10. Step values

Lets you specify the dimensional value (see Fig.9-23) of one of a series of regular consecutive cuts to progressive depths used in standard drill, deep and standard drill, breakchip cycles.

The software includes a step, S_1, S_2, $\cdots$, S_n parameter string in the cycle statement.

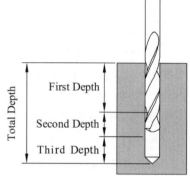

Fig.9-23　Step value

9.5　Drilling operation example

This example shows how to create different drilling operations using drilling geometry that is specified within the operation. The machined model is shown in Fig.9-24.

9.5.1　Creating a basic point-to-point drilling operation

(1) On the insert toolbar, click Create operation , or choose insert→operation.

Fig.9-24　Drilling operation model

(2) In the Create operation dialog box, from the type list, select drill. In the operation subtype group, click Drilling .

(3) On the insert toolbar, click Create tool , or choose insert→tool.

(4) In the Create tool dialog box, from the type list, select drill. In the tool subtype group, click

Drilling tool . The software displays the Drilling tool dialog box.

(5) In the Drilling tool dialog box, set the tool parameters as shown in Fig.9-25.

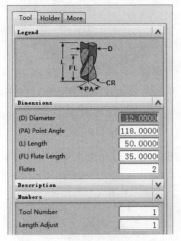

Fig.9-25 Creating drilling tool

(6) In the location group, set the options as shown in Tab.9-10.

Tab.9-10 Operation options

Program	Program
Tool	Drill_d12
Geometry	Workpiece
Method	Drill method

(7) Click OK. The Drilling dialog box appears.

(8) In the Geometry group, click Specify holes . The Point dialog box appears.

(9) Click Select. In the graphics window, select the edge of the holes successively as cycle parameter set 1 and cycle parameter set 2, as shown in Fig.9-26 and Fig.9-27.

Fig.9-26 Selecting the holes as set 1   Fig.9-27 Selecting the holes as set 2

(10) Click OK twice to confirm the hole selection, and return to the Drilling dialog box.

(11) In the Geometry group, click Specify part surface . The Part surface dialog box appears.

(12) In the graphics window, select the face of the part as shown in Fig.9-28.

Fig.9-28 Selecting the part surface

(13) In the geometry group, click Specify bottom surface . The Bottom surface dialog box appears.

(14) In the graphic window, select the bottom of the part as shown in Fig.9-29.

Fig.9-29 Selecting the bottom surface

(15) In the cycle type group, from the cycle list, select standard drill. in the cycle row, click Edit parameters . The Specify number of dialog box appears.

This example uses a standard drill cycle that contains two cycle parameter set. The cue line displays the number of cycle parameter sets you can use for this operation.

(16) Click OK to accept the default value of two for the cycle parameter set. The Cycle parameters dialog box appears.

(17) Set the cycle parameters for set 1 and set 2 successively, as shown in Fig.9-30 and Fig.9-31.

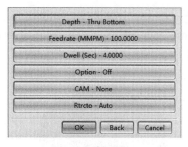 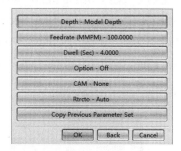

Fig.9-30 Cycle parameters for set 1   Fig.9-31 Cycle parameters for set 2

(18) Click OK twice to return to the Drilling dialog box.

(19) In the cycle type group, type in a minimum clearance value.

This example uses a value of 16. The minimum clearance value determines the distance above the material at which the tool is positioned before entering the hole.

(20) In the actions group, click Generate .

(21) Click OK to accept the tool path and close the dialog box. Drilling operation path shows in Fig.9-32.

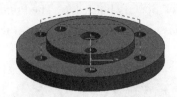

Fig.9-32 Drilling operation path

9.5.2 Creating a counterboring operation

(1) On the Insert toolbar, click Create operation , or choose insert→operation.

(2) In the Create operation dialog box, from the type list, select drill. In the operation subtype group, click Counterboring .

(3) On the Insert toolbar, click Create tool , or choose insert→tool.

(4) In the Create tool dialog box, from the type list, select drill. In the tool subtype group, click Counterboring tool . The software displays the Counterboring tool dialog box.

(5) In the Counterboring tool dialog box, set the tool parameters as shown in Fig.9-33.

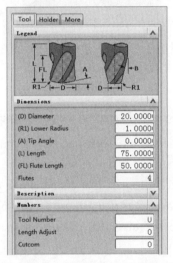

Fig.9-33 Creating counterboring tool

(6) In the Location group, set the options as shown in Tab.9-11.

Tab.9-11 Operation options

Program	Program
Tool	Counterboring_d20
Geometry	Workpiece
Method	Drill method

(7) Click OK. The Drilling dialog box appears.

223

(8) In the geometry group, click Specify holes . The Point dialog box appears.

(9) Click Select. In the graphics window, select the edge of the hole as shown in Fig.9-34.

Fig.9-34 Selecting the edge of the Hole

(10) Click OK twice to confirm the hole selection, and return to the Drilling dialog box.

(11) In the geometry group, click Specify part surface . The Part surface dialog box appears.

(12) In the graphics window, select the face of the part as shown in Fig.9-35.

Fig.9-35 Selecting the part surface

(13) In the cycle type group, from the cycle list, select standard drill. In the cycle row, click edit parameters . The Specify number of dialog box appears.

This example uses a standard drill cycle that contains one cycle parameter set. The cue line displays the number of cycle parameter sets you can use for this operation.

(14) Click OK to accept the default value of one for the cycle parameter set. The Cycle parameters dialog box appears.

(15) Set the cycle parameters, as shown in Fig.9-36.

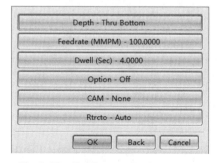

Fig.9-36 Setting cycle parameters

(16) Click OK twice to return to the Counterboring dialog box.

(17) In the Cycle type group, type in a minimum clearance value. This example uses a value of 3.

(18) In the actions group, click Generate .

(19) Click OK to accept the tool path and close the dialog box. Fig.9-37 shows the counterboring operation path.

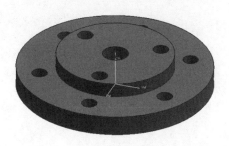

Fig.9-37 Counterboring operation path

Notes

[1] For multiple holes of the same diameter in a part, you can create the drilling operations rapidly by specifying different drilling cycles and cycle parameters rather than specifying each hole individually, which will make the tool changing and repositioning unnecessarily, so as to save considerable machining time and improve machining efficiency.

当零件中出现多个直径相同的孔时，可通过指定不同的循环方式和循环参数组进行加工，而不需要分别指定每个孔。这样就不需要换刀和重定位，从而减少了加工时间，提高了加工效率。

[2] Fourteen drill operation subtypes are available, as shown in Tab.9-1, which all use the drill point-to-point processor for removing material when manufacturing holes, with the exception of mill control, mill user, and thread milling.

钻孔加工一共有14种子类型，如表9-1所示。除了切削控制、用户自定义和螺纹加工外，其他子类型在加工孔时，均采用点到点钻孔处理器来去除材料。

[3] For the peck drill and break chip simulated cycles, if you specify a dwell value in seconds or revolutions, the software generates a DELAY/t or DELAY/rev, r commands statement to activate the desired dwell after the tool has been fed to depth.

对于啄钻和段屑循环，如果定义了停车时间（秒或者转数），在刀具进给到指定深度之后，软件将产生一个延迟 t 或 REV 以及 r 命令，来激活预期的停车时间。

New Words in Chapter 9

dwell	[dwel]	n.	停歇，停车
break chip		n.	断屑
through-hole		n.	通孔，金属化孔

blind hole		n.	盲孔；闷眼；不通孔
counterbore	['kaʊntə,bɔː]	n.	平底埋头孔，扩孔
countersink	['kaʊntəsɪŋk]	n.	倒角沉头孔
spot-facing		n.	锪孔，扩孔
peck-drilling		n.	啄钻
bore	[bɔː]	vi.	钻孔，镗孔
ream	[riːm]	vi.	铰孔
tap	[tæp]	vi.	攻螺纹
thread	[θred]	n.	螺纹

Chapter 10
A Comprehensive CAM Instance

Objectives:
✓ To understand the typical flow of a CAM system.
✓ To complete the milling operation of a moderately complicated part.
✓ To be able to postprocess an operation and get the output file.

10.1 Part analysis

This example shows how to complete the milling operations of a moderately complicated part, which is shown in Fig.10-1.

Before creating the manufacturing plan, the part should be analyzed to collect the relevant data, which include:

· Information to specify the correct cutting tools and tool parameters.

· Information to specify the correct stock requirements, such as the material type, and required tolerances.

Fig.10-1 A moderately complicated part

The size of the part can be obtained through analysis→measure distance, as shown in Fig.10-2.

Fig.10-2 Measure the part size

You can analyze the other information of the part in the same way or with NC assistant (see Fig.10-3).

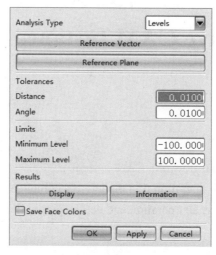

Fig.10-3 NC assistant Dialog Box

After the part analysis has been completed, the manufacturing plan for the part can be drawn up, as shown in Tab.10-1.

Tab.10-1 The part manufacturing plan

NO	Operation	Subtype	Method	Machine tool	Tolerance	Stock
1	Machining the entire part	Cavity milling	Rough	Mill_D20R4	0.05	0.35
2	Machining the part with IPW	Rest milling	Rough	Mill_D16R2	0.05	0.2
3	Machining the plan on the part	Face milling	Semi-finish	Mill_D16R2	0.03	0.1
4	Machining the steep on the part	Z-level	Semi-finish	Mill_D12R2	0.03	0.1
5	Machining the small cavity	Fixed contour	Semi-finish	Mill_D10R5	0.03	0.1
6	Machining the surface corners	Fixed contour	Semi-finish	Mill_D12R6	0.03	0.1
7	Machining the hemispherical cavity	Fixed contour	Semi-finish	Mill_D10R5	0.03	0.1
8	Machining the entire surface	Fixed contour	Semi-finish	Mill_D12R6	0.03	0.1
9	Finishing the plan on the part	Face milling	Finish	Mill_D16R2	0.01	0
10	Finishing the side wall and corners	Fixed contour	Finish	Mill_D4R2	0.01	0
11	Finishing the small cavity	Fixed contour	Finish	Mill_D4R2	0.01	0
12	Finishing the hemispherical cavity	Fixed contour	Finish	Mill_D10R5	0.01	0
13	Finishing the steep on the part	Z-level	Finish	Mill_D12R2	0.01	0
14	Finishing the entire surface	Fixed contour	Finish	Mill_D10R5	0.01	0

10.2 Setting up the machining environment

The following steps show how to set up the machining environment.
(1) Choose start→manufacturing.
(2) In the Machining environment dialog box, in the CAM session configuration group, select *cam general* from the list.
(3) In the CAM setup to create group, select *mill contour* from the list as shown in Fig.10-4.
(4) Click OK.

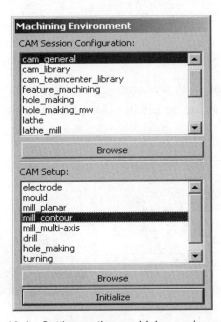

Fig.10-4　Setting up the machining environment

10.3 Preparation for creating operations

10.3.1 Create an MCS geometry parent group

The MCS determines the orientation and origin of tool paths for all operations in the orient group. The MCS geometry parent group can be created as follows:
(1) On the insert toolbar, click Create geometry , or choose insert→geometry.
(2) In the Create geometry dialog box, do the following as shown in Fig.10-5:
① In the geometry subtype group, click MCS .
② In the location group, select geometry to assign your new MCS.
③ In the name group, enter *MCS001* for the new MCS, and click OK.

229

Fig.10-5   Create geometry dialog box

(3) In the MCS dialog box, the MCS group, click CSYS dialog .

(4) In the CSYS dialog box, click Dynamic .

(5) In the graphics window, use the handles on the MCS to move it to the top left corner of the part (see Fig.10-6).

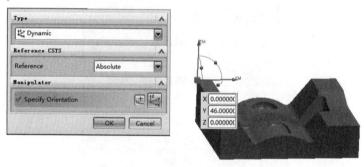

Fig.10-6   Define MCS

(6) Click OK to return to the MCS dialog box.

(7) In the clearance group, select plane as the clearance option.

(8) Click specify clearance plane.

(9) In the Plane constructor dialog box, define the top face of the part for the clearance plane, and enter *5.0* as offset distance (as shown in Fig.10-7).

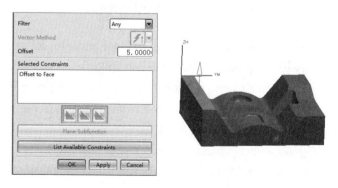

Fig.10-7   Define clearance plane

(10) Click OK to save and close the dialog box.

(11) In the MCS dialog box, click OK.

10.3.2　Create geometry

(1) On the insert toolbar, click Create geometry, or choose insert→geometry.

(2) In the Create geometry dialog box, select Mill contour from the type list.

(3) In the geometry subtype group, click Workpiece.

(4) In the location group, select MCS001 from the geometry list. The new geometry parent inherits geometry from the location you select.

(5) In the name group, enter mill geom001 for the new geometry.

(6) Click OK. The Workpiece dialog box opens as shown in Fig.10-8.

Fig.10-8　Workpiece dialog box

(7) In the Workpiece dialog box, in the geometry group, click Specify part.

(8) The Part geometry dialog box opens.

(9) In the graphics window, select the entire part as part geometry (see Fig.10-9).

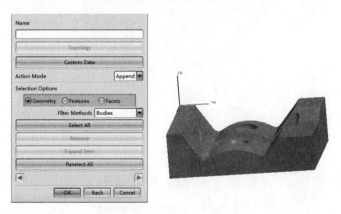

Fig.10-9　Define part geometry

(10) Click OK to close the Part geometry dialog box.

(11) In the Workpiece dialog box, click Specify blank ⊗. The Blank geometry dialog box opens.

(12) In the selection options group , select Auto Block radio box. The NX creates a block containing the model automatically, as shown in Fig.10-10.

(13) Click OK to close the Blank geometry dialog box.

(14) Click OK. The mill geom001 has been specified.

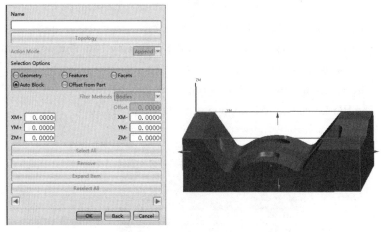

Fig.10-10　Define blank geometry

10.3.3　Create machine tool

(1) On the insert toolbar, click Create tool 🛠, or choose insert→tool as shown in Fig.10-11.

(2) In the Create tool dialog box, from the type list, select *Mill_contour*. In the tool subtype group, click Mill 🛠. Set machine tool name as Mill_D20R4.

Fig.10-11　Create tool dialog

(3) The software displays the Five parameter mill tool dialog box.

(4) In the Five parameter mill tool dialog box, set the tool parameters as shown in Fig10-12.

232

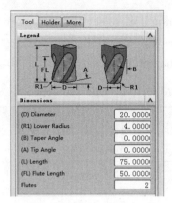

Fig.10-12 Define tool parameters

(5) Click OK. Mill_D20R4 is created.

(6) Create the other 6 machine tools in Tab.10-1 in the same way.

10.3.4 Create machine method

(1) In the machine method view (see Fig.10-13), double click Mill_rough, opens Mill_method dialog box.

(2) In the Mill method dialog box, set part stock value as 0.35.

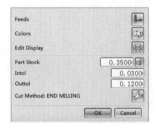

Fig.10-13 Set machine method

(3) Click OK.
(4) Double click Mill semi-finish, opens Mill method dialog box.
(5) Set part stock value as 0.1.
(6) Double click Mill finish, opens mill method dialog box.
(7) Set part stock value as 0.

10.4 Create milling operations

10.4.1 Create cavity milling

(1) On the insert toolbar, click Create operation or choose insert→operation.

(2) In the Create operation dialog box, select mill_contour from the type List. In the operation Subtype group, click Cavity milling.

(3) In the location group, set the options as shown in Tab.10-2.

Tab.10-2　Operation options

Program	Program
Tool	Mill_d20r4
Geometry	Mill_geom001
Method	Mill_rough

(4) In the name group, enter Cavity_mill_01 for the new operation.

(5) Click OK. The Cavity mill dialog box is displayed.

(6) In the geometry group, click Specify trim boundaries (see Fig.10-14). Opens Trim boundaries dialog.

(7) In the Trim boundaries dialog, select face boundary from filter type.

(8) In the graphic windows, select the bottom face of the part.

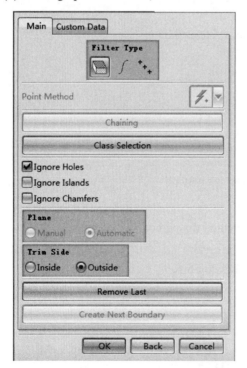
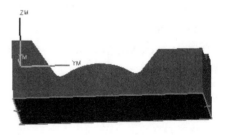

Fig.10-14　Specify trim boundaries

(9) In the trim side group, select outside radio box.

(10) Click OK to return to the Cavity milling dialog.

(11) In the path setting group, set the options as shown in Tab.10-3.

Tab.10-3 Path setting options

Cutter pattern		Follow periphery
Stepover		% Tool flat
Percent of flat diameter		65.0
Global depth per cut		2
Feeds and speeds	Spindle speed	2200
	Cut	800
	Rapid	2500
	Approach	0
	Engage	400
	First cut	400
	Step over	600
	Traversal	1500
	Retract	1500
	Departure	0

(12) In the path setting group, click Cut levels ![icon]. Opens Cut levels dialog box as shown in Fig.10-15.

(13) In the range definition group, click Downward ![icon] continuously, till range 4 is the current range.

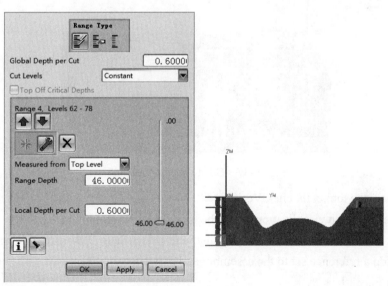

Fig.10-15 Edit cut levels

(14) Click Delete current range ![X].

235

(15) Click OK to close the Cut levels dialog.
(16) Click Cutting parameters ⬚, opens the Cutting parameters dialog box.
(17) In the containment tab, select *Use 3D* from IPW list in the blank group (see Fig.10-16).
(18) Click OK to return to the Cavity milling dialog box.

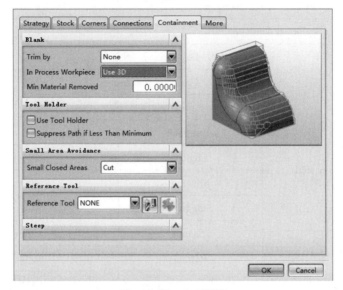

Fig.10-16 Set IPW

(19) In the Cavity milling dialog box, in the actions group, click Generate ⬚ as shown in Fig.10-17.

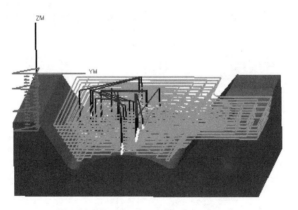

Fig.10-17 Cavity milling tool path

(20) To view the tool path, click Verify ⬚.
(21) In the Tool path visualization dialog box, click the 2D dynamic tab.
(22) Select medium from generate IPW list, and select Save IPW as component check box. The NX will Add a reference set to the assembly and create a component part.
(23) Click Play ▶.
(24) Click OK to close the Tool path visualization dialog box.
(25) In the actions group, click Display resulting IPW ⬚, the IPW is shown as Fig.10-18.

236

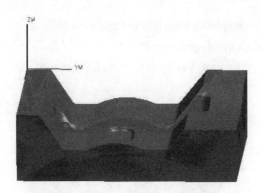

Fig.10-18 Display resulting IPW

(26) Click OK to save the operation and close the Cavity milling dialog box.

10.4.2 Create rest milling with IPW

(1) On the insert toolbar, click Create operation or choose insert→operation.

(2) In the Create operation dialog box, select Mill contour from the type List. In the operation subtype group, click Rest milling .

(3) In the location group, set the options as shown in Tab.10-4.

Tab.10-4 Operation options

Program	Program
Tool	Mill_d16r2
Geometry	Mill_geom001
Method	Mill_rough

(4) In the name group, enter Rest_milling_01 for the new operation.

(5) Click OK. The Rest milling dialog box is displayed.

(6) In the geometry group, click Specify trim boundaries in the same way with the steps of 10.4.1.

(7) In the path setting group, set the options as shown in Tab.10-5.

Tab.10-5 Path setting options

Cutter pattern		Follow periphery
Stepover		% Tool flat
Percent of flat diameter		65.0
Global depth per cut		1
Cutting parameters	Stock-Part side stock	0.2
	Containment-IPW	Use 3D
Feeds and speeds		As shown in Tab.10-3

237

(8) In the geometry group, you can display the previous IPW.

(9) In the actions group, click Generate .

(10) To view the tool path, click Verify . Rest milling tool path is shown in Fig.10-19.

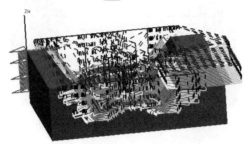

Fig.10-19 Rest milling tool path

(11) Click OK to save the operation and close the Rest milling dialog box.

10.4.3 Create planar milling with semi-finish method

(1) On the insert toolbar, click create operation or choose insert→operation.

(2) In the Create operation dialog box, in the type group, click Mill planar.

(3) In the operation subtype group, click Face milling .

(4) In the location group, set the options as shown in Tab.10-6.

Tab.10-6 Operation options

Program	Program
Tool	Mill_d16r2
Geometry	Mill_geom001
Method	Mill_semi-finish

(5) In the name group, enter Face_milling_01 for the new operation.

(6) Click OK. The Face milling dialog box is displayed.

(7) In the geometry group, click Specify face boundaries , opens Specify face geometry dialog as shown in Fig.10-20.

Fig.10-20 Specify face geometry

238

(8) Select *face boundary* from filter type.

(9) In the graphic windows, select the three faces of the part, as shown in Fig.10-20.

(10) Click OK to return to the Face milling dialog.

(11) In the path setting group, set the options as shown in Tab.10-7.

Tab.10-7　Path setting options

Cutter pattern	Follow periphery
Stepover	% Tool flat
Percent of flat diameter	50.0
Blank distance	0.4
Depth per cut	0.25
Final floor stock	0.1
Feeds and speeds	As shown in Tab.10-3

(12) In the actions group, click Generate .

(13) To view the tool path, click Verify as shown in Fig.10-21.

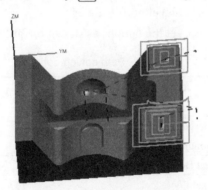

Fig.10-21　Face milling tool path

(14) Click OK to save the operation and close the Z-level profile dialog box.

10.4.4　Create Z-level milling with semi-finish method

(1) On the insert toolbar, click Create operation , or choose insert→operation.

(2) In the Create operation dialog box, from the type list, select mill contour.

(3) In the operation subtype group, click Z-level profile .

(4) In the location group, set the options as shown in Tab.10-8.

Tab.10-8　Operation options

Program	Program
Tool	Mill_d16r2
Geometry	Mill_geom001
Method	Mill_semi-finish

(5) In the name group, enter Z-level_profile_01 for the new operation.

(6) Click OK. The Z-level profile dialog box is displayed.

(7) In the geometry group, click specify cut area . Opens Cut area dialog.

(8) In the graphic windows, select the side wall of the part, as shown in Fig.10-22.

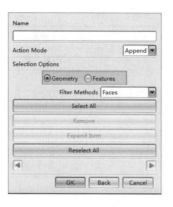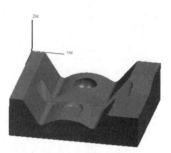

Fig.10-22   Specify cut area

(9) Click OK to return to the Z-level profile dialog.

(10) In the path setting group, set the options as shown in Tab.10-9.

Tab.10-9   Path setting options

	Steep containment	Steep only
	Angle	65
	Merge distance	3.0
	Minimum cut length	1.0
	Global Depth per Cut	0.2
	Final floor stock	0.1
Cutting parameters	Strategy-cut order	Level first
	Strategy-roll tool over edges	On
	Feeds and speeds	As shown in Tab.10-3

(11) In the actions group, click Generate .

(12) To view the tool path, click Verify . The tool path is shown in Fig.10-23.

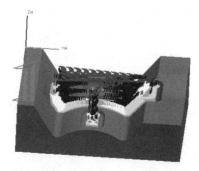

Fig.10-23   Z-level milling tool path

(13) Click OK to save the operation and close the Face milling dialog box.

10.4.5 Create fixed contour milling with area milling drive method

(1) On the insert toolbar, click Create operation , or choose insert→operation.
(2) In the Create operation dialog box, from the type list, select mill_contour.
(3) In the operation subtype group, click Fixed_contour .
(4) In the Location group, set the options as shown in Tab.10-10.

Tab.10-10 Operation options

Program	Program
Tool	Mill_d10r5
Geometry	Mill_geom001
Method	Mill_semi-finish

(5) In the Name group, enter Fixed_contour_01 for the new operation.
(6) Click OK. The Fixed contour dialog box is displayed.
(7) In the geometry group, click Specify cut area . Opens Cut area dialog.
(8) In the graphic windows, select the bottom face of the small cavity, as shown in Fig.10-24.

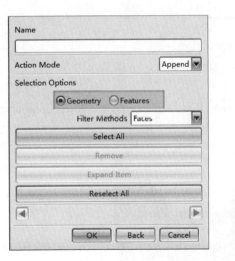 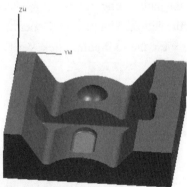

Fig.10-24 Specify cut area

(9) Click OK to return to the Fixed contour dialog.
(10) In the drive method group, select area milling from the method list. Opens area milling drive method dialog as shown in Fig.10-25.

Fig.10-25　Area milling drive method dialog

(11) In the Area milling drive method dialog, set the options as shown in Tab.10-11.

Tab.10-11　Area milling drive method options

Steep containment method	None
Cut pattern	Follow periphery
Pattern direction	Inward
Cut direction	Conventional cut
Stepover	Constant
Distance	0.1
Stepover applied	On plane

(12) Click OK to return to the Fixed contour dialog.

(13) In the path setting group, set feeds and speeds as shown in Tab.10-3.

(14) In the actions group, click Generate .

(15) To view the tool path, click Verify . Fig.10-26 shows the fixed contour milling tool path.

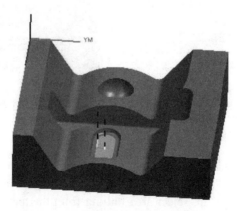

Fig.10-26　Fixed contour milling tool path with area milling drive method

(16) Click OK to save the operation and close the Fixed contour dialog box.

242

10.4.6 Create fixed contour milling with flow cut drive method

(1) On the insert toolbar, click Create operation, or choose insert→operation.
(2) In the Create operation dialog box, from the type list, select mill_contour.
(3) In the operation subtype group, click Fixed contour.
(4) In the location group, set the options as shown in Tab.10-12.

Tab.10-12 Operation options

Program	Program
Tool	Mill_d12r6
Geometry	Mill_geom001
Method	Mill_semi-finish

(5) In the name group, enter Fixed_contour_02 for the new operation.
(6) Click OK. The Fixed contour dialog box is displayed.
(7) In the geometry group, click Specify cut area. Opens Cut area dialog.
(8) In the graphic windows, select the surface of the part, as shown in Fig.10-27.

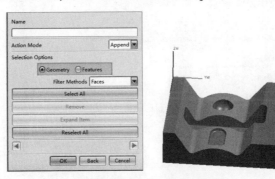

Fig.10-27 Specify cut area

(9) Click OK to return to the Fixed contour dialog.
(10) In the drive method group, select flow cut from the method list. Opens Flow cut drive method dialog as shown in Fig.10-28.

Fig.10-28 Flow cut drive method dialog

(11) In the Area milling drive method dialog, set the options as shown in Tab.10-13.

Tab.10-13 Flow cut drive method options

Max concavity	179
Minimum cut length	1.0
Merge distance	3.0
Flowcut type	Reference tool offsets
Cut pattern	Zig-zag
Stepover	1.5
Sequencing	Inside-out
Reference tool diameter	20
Overlap distance	2

(12) In the path setting group, Click Cutting parameters . The Cutting parameters dialog box appears as shown in Fig.10-29.

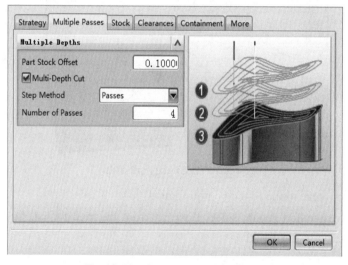

Fig.10-29 Set cutting parameters

(13) In the Cutting parameters dialog box, click the Multiple passes tab. Type 0.1 in the Part stock offset box.

(14) Select Multi-depth cut check box.

(15) In the step method list, select passes. Type 4 in the Number of passes input box.

(16) Click OK to return to the Fixed contour dialog.

(17) In the path setting group, set feeds and speeds as shown in Tab.10-3.

(18) In the actions group, click Generate .

(19) To view the tool path (see Fig.10-30), click Verify .

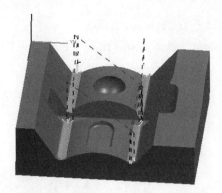

Fig.10-30 Fixed contour milling tool path with flow cut drive method

(20) Click OK to save the operation and close the Fixed contour dialog box.

10.4.7 Create fixed contour milling with spiral drive method

(1) On the insert toolbar, click Create operation, or choose insert→operation.

(2) In the Create operation dialog box, from the type list, select mill contour.

(3) In the operation subtype group, click Fixed_contour.

(4) In the location group, set the options as shown in Tab.10-14.

Tab.10-14 Operation options

Program	Program
Tool	Mill_d10r5
Geometry	Mill_geom001
Method	Mill_semi-finish

(5) In the name group, enter Fixed_contour 03 for the new operation.

(6) Click OK. The Fixed contour dialog box is displayed.

(7) In the geometry group, click Specify cut area. Opens Cut area dialog.

(8) In the graphic windows, select the hemispherical cavity of the part, as shown in Fig.10-31.

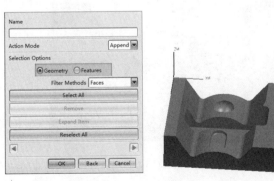

Fig.10-31 Specify cut area

(9) Click OK to return to the Fixed contour dialog.

(10) In the drive method group, select *spiral* from the method list. Opens Spiral drive method dialog as shown in Fig.10-32.

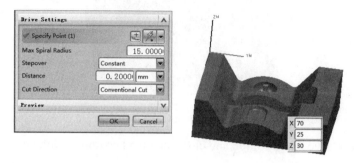

Fig.10-32 Spiral drive method dialog

(11) In the Spiral drive method dialog, click Specify point. Select the center point of the hemispherical cavity.

(12) In the drive setting group, set the options as shown in Tab.10-15.

Tab.10-15 Spiral drive method options

Max spiral radius	15
Stepover	Constant
Distance	0.2
Cut direction	Conventional cut

(13) Click OK to return to the Fixed contour dialog.

(14) In the path setting group, set feeds and speeds as shown in Tab.10-3.

(15) In the actions group, click Generate.

(16) To view the tool path (see Fig.10-33), click Verify.

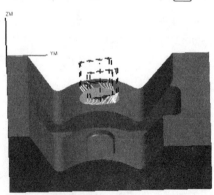

Fig.10-33 Fixed Contour Milling Tool Path with spiral drive method

(17) Click OK to save the operation and close the Fixed contour dialog box.

10.4.8 Create fixed contour milling with area milling drive method for the entire surface

(1) On the insert toolbar, click Create operation , or choose insert→operation.
(2) In the Create operation dialog box, from the type list, select mill contour.
(3) In the operation subtype group, click Fixed contour .
(4) In the location group, set the options as shown in Tab.10-16.

Tab.10-16 Operation options

Program	Program
Tool	Mill_d12r6
Geometry	Mill_geom001
Method	Mill_semi-finish

(5) In the name group, enter Fixed_contour_04 for the new operation.
(6) Click OK. The Fixed contour dialog box is displayed.
(7) In the geometry group, click Specify cut area . Opens Cut area dialog.
(8) In the graphic windows, select the bottom face of the small cavity, as shown in Fig.10-34.

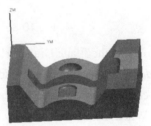

Fig.10-34 Specify cut area

(9) Click OK to return to the Fixed contour dialog.
(10) In the drive method group, select area milling from the method list. Opens area milling drive method dialog.
(11) In the Area milling drive method dialog, set the options as shown in Tab.10-17.

Tab.10-17 Area milling drive method options

Steep containment method	None
Cut pattern	Zig-zag
Cut direction	Conventional cut
Stepover	Constant

continue

Steep containment method	None
Distance	0.3
Stepover applied	On plane
Cut angle	User defined
Degrees	45.0

(12) Click OK to return to the Fixed contour dialog.

(13) In the geometry group, click Specify trim boundaries . Opens Trim boundary dialog.

(14) In the Trim boundary dialog, select *curve boundary* of the filter type group.

(15) In the plane group, select Manual radio box. Opens Plane dialog (see Fig.10-35).

(16) In the plane dialog, click ZC Constance, and type 0.0 in the input box.

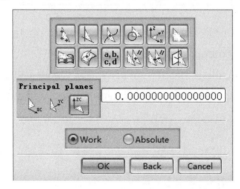

Fig.10-35 Specify plane

(17) Click OK to close the Plane Dialog.

(18) Select *Inside* radio box in the Trim Side group.

(19) Select the curve boundaries of the cavity in sequence, which has been machined in the previous operations. When the curve boundary is selected, click Create next boundary as shown in Fig.10-36.

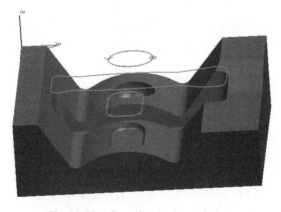

Fig.10-36 Specify trim boundaries

(20) Click OK to return to the Fixed contour dialog.
(21) In the path setting group, set feeds and speeds as shown in Tab.10-3.
(22) In the Actions group, click Generate .
(23) To view the tool path (see Fig.10-37), click Verify .

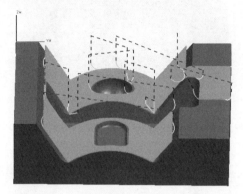

Fig.10-37 Fixed contour milling tool path

(24) Click OK to save the operation and close the Fixed contour dialog box.

10.4.9 Create planar milling with finish method

The planar milling operation with finish Method is similar with the semi-finish operation, so you can copy the previous planar milling operation.

(1) In the program order view, copy Face_milling_01 node, and paste it under the program node.
(2) Rename the new operation as Face_milling_02.
(3) Double click Face_milling_02, opens Face milling dialog.
(4) In the path setting group, select mill finish from the method list.
(5) In the path setting group, set the options as shown in Tab.10-18.

Tab.10-18 Path setting options

Cutter pattern	Follow periphery
Stepover	% Tool flat
Percent of flat diameter	50.0
Blank distance	0.15
Depth per cut	0.15
Final floor stock	0.0
Feeds and speeds	As shown in Tab.10-3

(6) In the actions group, click Generate .
(7) To view the tool path (see Fig.10-38), click Verify .

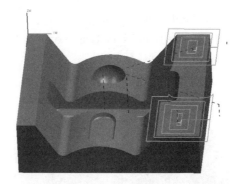

Fig.10-38   Planer milling tool path with finish method

(8) Click OK to save the operation and close the Face milling dialog box.

10.4.10   Create finish fixed contour milling with radial cut drive method

(1) On the insert toolbar, click Create operation , or choose insert→operation.
(2) In the Create operation dialog box, from the type list, select mill contour.
(3) In the operation subtype group, click Fixed contour .
(4) In the location group, set the options as shown in Tab.10-19.

Tab.10-19   Operation options

Program	Program
Tool	Mill_d4r2
Geometry	Mill_geom001
Method	Mill_finish

(5) In the name group, enter Fixed_contour_05 for the new operation.
(6) Click OK. The Fixed contour dialog box is displayed.
(7) In the drive method group, select radial cut from method list. Opens Radial cut drive method dialog (see Fig.10-39).

Fig.10-39   Radial cut drive method dialog

(8) In the drive geometry group, click Specify drive geometry, Opens Temporary boundary dialog (see Fig.10-40).

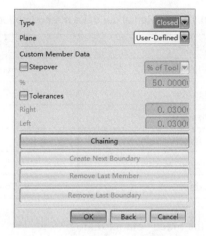

Fig.10-40 Temporary boundary dialog

(9) Select *closed* from type list.

(10) Select *user-defined* from plane list, opens Plane dialog.

(11) In the Plane dialog, click ZC constance, and type 0.0 in the input box as shown in Fig.10-41.

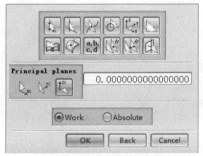

Fig.10-41 Specify plane

(12) Click OK to close the Plane dialog.

(13) Select the boundary of the small cavity, as shown in Fig.10-42.

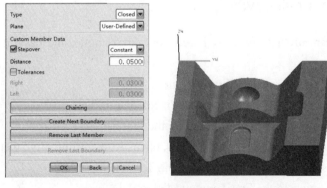

Fig.10-42 Create boundary

251

(14) Click OK to return to the Radial cut drive method dialog.

(15) In the Drive setting group, set the options as shown in Tab.10-20.

Tab.10-20 Radial cut drive method options

Cut type	None
Cut direction	Conventional cut
Stepover	Constant
Distance	0.1
Band on material side	5.0
Band on opposite side	2.0
Path direction	Follow boundary

(16) Click OK to return to the Fixed contour dialog.

(17) In the Path setting group, set feeds and speeds as shown in Tab.10-3.

(18) In the actions group, click Generate .

(19) To view the tool path (see Fig.10-43), click Verify .

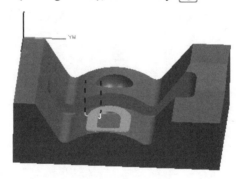

Fig.10-43 Fixed contour milling tool path with radial cut drive method

(20) Click OK to save the operation and close the Fixed contour dialog box.

10.4.11 Create finish fixed contour milling with area milling drive method for the small cavity

(1) In the program order view, copy Fixed_contour_01 node, and paste it under the program node.

(2) Rename the new operation as Fixed_contour_06.

(3) Double click Fixed_contour_06, opens Fixed contour dialog.

(4) In the tool group, select Mill_D4R2 from the tool list.

(5) In the drive method group, click Edit , opens Area milling drive method dialog.

(6) In the drive settings group, type 0.05 in the distance input box.

(7) Click OK to return to the Fixed contour dialog.

(8) In the path setting group, select Mill finish from the method list.
(9) In the actions group, click Generate.
(10) To view the tool path (see Fig.10-44), click Verify.

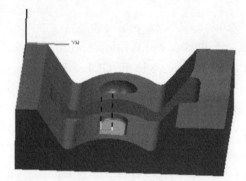

Fig.10-44　Finish fixed contour milling tool path with area milling drive method

(11) Click OK to save the operation and close the Fixed contour dialog box.

10.4.12　Create finish fixed contour milling with spiral drive method

(1) In the Program order view, copy Fixed_contour_03 node, and paste it under the program node.
(2) Rename the new operation as Fixed_contour_07.
(3) Double click Fixed_contour_07, opens Fixed contour dialog.
(4) In the tool group, select Mill_D4R2 from the Tool list.
(5) In the drive method group, click Edit, opens Spiral drive method dialog.
(6) In the drive settings group, type 0.05 in the distance input box.
(7) Click OK to return to the Fixed contour dialog.
(8) In the path setting group, select Mill finish from the method list.
(9) In the actions group, click Generate.
(10) To view the tool path, click Verify. The tool path is shown in Fig.10-45.
(11) Click OK to save the operation and close the Fixed contour dialog box.

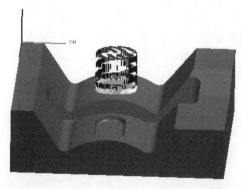

Fig.10-45　Finish fixed contour milling tool path with spiral drive method

10.4.13 Create finish Z-level milling

(1) In the program order view, copy Z-level_profile_01 node, and paste it under the program node.

(2) Rename the new operation as Z-level_profile_02.

(3) Double click Z-level_profile_02., opens Z-level profile dialog.

(4) In the drive settings group, type 0.08 in the Global depth per cut input box.

(5) Click Cutting parameters , opens the Cutting parameters dialog box.

(6) In the strategy Tab, clear Roll tool over edges check box.

(7) Click OK to return to the Z-level profile dialog.

(8) In the actions group, click Generate .

(9) To view the tool path (see Fig.10-46), click Verify .

(10) Click OK to save the operation and close the Z-level profile dialog box.

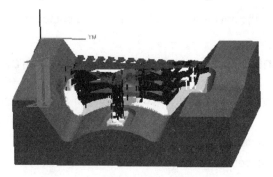

Fig.10-46　Finish Z-level milling tool path

10.4.14 Create finish fixed contour milling with area milling drive method for the entire surface

(1) In the program order view, copy Fixed_contour_04 node, and paste it under the program node.

(2) Rename the new operation as Fixed_contour_08.

(3) Double click Fixed_contour_08, opens Fixed contour dialog.

(4) In the drive method group, click Edit , opens Area milling drive method dialog.

(5) In the drive settings group, type 0.1 in the Distance input box.

(6) Type 135.0 in the degrees input box.

(7) Click OK to return to the Fixed contour dialog.

(8) In the path setting group, select Mill finish from the method list.

(9) In the actions group, click Generate .

(10) To view the tool path (see Fig.10-47), click Verify .

(11) Click OK to save the operation and close the Fixed contour dialog box.

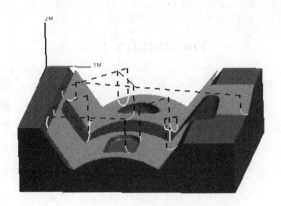

Fig.10-47 Finish fixed contour milling tool path with area milling drive method

Since then the CAM operations of the part have been complete. The operations are shown in machine method view and machine tool view, as shown in Fig.10-48.

Machine Tool View			Machine Method View		
⊖ MILL_D20R4			⊖ MILL_ROUGH		
CAVITY_MILL_01	✓	MILL_D20R4	CAVITY_MILL_01	✓	MILL_D20R4
⊖ MILL_D16R2			REST_MILLING_01	✓	MILL_D16R2
REST_MILLING_01	✓	MILL_D16R2	⊖ MILL_SEMI_FINISH		
FACE_MILLING_01	✓	MILL_D16R2	ZLEVEL_PROFILE_01	✓	MILL_D12R2
FACE_MILLING_02	✓	MILL_D16R2	FIXED_CONTOUR_01	✓	MILL_D10R5
⊖ MILL_D12R2			FIXED_CONTOUR_02	✓	MILL_D12R6
ZLEVEL_PROFILE_01	✓	MILL_D12R2	FIXED_CONTOUR_03	✓	MILL_D10R5
ZLEVEL_PROFILE_02	✓	MILL_D12R2	FACE_MILLING_01	✓	MILL_D16R2
⊖ MILL_D12R6			FIXED_CONTOUR_06	✓	MILL_D4R2
FIXED_CONTOUR_02	✓	MILL_D12R6	FIXED_CONTOUR_04	✓	MILL_D12R6
FIXED_CONTOUR_08	✓	MILL_D12R6	⊖ MILL_FINISH		
FIXED_CONTOUR_04	✓	MILL_D12R6	FACE_MILLING_02	✓	MILL_D16R2
⊖ MILL_D10R5			FIXED_CONTOUR_05	✓	MILL_D4R2
FIXED_CONTOUR_01	✓	MILL_D10R5	FIXED_CONTOUR_07	✓	MILL_D4R2
FIXED_CONTOUR_03	✓	MILL_D10R5	ZLEVEL_PROFILE_02	✓	MILL_D12R2
⊖ MILL_D4R2			FIXED_CONTOUR_08	✓	MILL_D12R6
FIXED_CONTOUR_05	✓	MILL_D4R2			
FIXED_CONTOUR_06	✓	MILL_D4R2			
FIXED_CONTOUR_07	✓	MILL_D4R2			

Fig.10-48 Operation in the machine tool view and machine method view

Vocabulary List

Absolute Coordinate System (ACS)			绝对坐标系
accuracy	[ˈækjʊrəsɪ]	n.	精确度，准确性
Alternating Current(AC)			交流电
armature	[ˈɑːmətʃə; -tj(ʊ)ə]	n.	电枢
assembly	[əˈsemblɪ]	n.	装配，装配体
associativity	[əˌsəʊʃjəˈtɪvəti]	n.	相关性
avoidance	[əˈvɒɪdəns]	n.	避让
backdraft	[ˈbækˌdræft]	n.	倒转，回程
backlash	[ˈbæklæʃ]	n.	反冲
bandwidth	[ˈbændwɪdθ]	n.	带宽
Basic Length Unit(BLU)			基本长度单元
blend	[blend]	n.	圆角
blind hole		n.	盲孔；闷眼；不通孔
bore	[bɔː]	vi.	钻孔，镗孔
break chip		n.	断屑
breakage	[ˈbreɪkɪdʒ]	n.	破坏，破损
B-surface			B 级表面
cable	[ˈkeɪb(ə)l]	n.	电缆
calibration	[kælɪˈbreɪʃ(ə)n]	n.	校准；刻度；标度
casting	[ˈkɑːstɪŋ]	n.	铸造，铸件
cavity milling			型腔铣
chamfer	[ˈtʃæmfə]	n.	斜面，凹槽
chip	[tʃɪp]	n.	切屑，凿
clamp	[klæmp]	n.	夹钳，螺丝钳
clearance plane			安全平面
climb cut			顺铣
clockwise	[ˈklɒkwaɪz]	adj.	顺时针的
closed-loop	[ˈkləʊzdluːp]	adj.	闭合环路的，闭环的
Computer Aided Manufacturing (CAM)			计算机辅助制造
concavity	[kɒnˈkævəti]	n.	凹度，凹角
concentric	[kənˈsentrɪk]	adj.	同轴的，同中心的
constant	[ˈkɒnst(ə)nt]	n.	常数；恒量

contour	['kɒntʊə]	n.	轮廓
contour milling			轮廓铣
Control Loop Unit (CLU)			环路控制器
conventional cut			逆铣
convex	['kɒnveks]	n.	凸面，凸状
coolant	['kuːl(ə)nt]	n.	冷却剂
counterbore	['kaʊntəˌbɔː]	n.	平底埋头孔，扩孔
counter-clockwise			逆时针的
countersink	['kaʊntəsɪŋk]	n.	倒角沉头孔
curvature	['kɜːvətʃə]	n.	曲率
Cutter Location Source File (CLSF)			刀位源文件
Data Processing Unit (DPU)			数据处理器
Direct Current (DC)			直流电
Distributed Numerical Control (DNC)			分布式数字控制
draft	[dræft]	n.	拔模
dwell	[dwel]	n.	停歇，停车
Electro Discharge Machine (EDM)			电火花加工机床
Electronic Industry Association (EIA)			电子工业协会
enclose	[ɪn'kloz]	vt.	围绕
encoder	[en'kəʊdə]	n.	编码器
engage	[ɪn'ɡeɪdʒ]	vi.	进刀
Ethernet	['iːθənet]	n.	以太网
face milling			面铣削
facet	['fæsɪt]	n.	面，小平面
feature-based machining (FBM)			基于特征加工
feedback	['fiːdbæk]	n.	反馈
finish	['fɪnɪʃ]	vi.	精加工
fixed-axis surface contouring			固定轴表面轮廓铣
flexible automation			柔性自动化
flow cut			清根
fluid	['fluːɪd]	n.	流体，液体
follow part			跟随工件
follow periphery			跟随周边
gear	[ɡɪə]	n.	齿轮，传动装置
gouge	[ɡaʊdʒ]	n.	沟，圆凿
Graphics Postprocessor Module (GPM)			图形后置处理模块

groove	[gruːv]	n.	凹槽，沟槽
helical	[ˈhelɪkəl]	adj.	螺旋形的
hierarchical	[haɪəˈrɑːkɪk(ə)l]	adj.	分层的
hierarchy	[ˈhaɪərɑːkɪ]	n.	层级
hookup	[ˈhʊkˌʌp]	n.	连接
hydraulic	[haɪˈdrɔlɪk]	adj.	液压的，水力的
In Process Workpiece (IPW)			处理中工件
Initial Graphic Exchange Specification (IGES)			初始图形交换规范
interpolator	[ɪnˈtəːpəʊleɪtə]		插补器
intersect	[ɪntəˈsekt]	vi.	相交，交叉
isolate	[ˈaɪsəleɪt]	adj.	隔离的，孤立的
Machine Control Unit (MCU)			机床控制器
Machine Coordinate System (MCS)			加工坐标系
magnetic	[mægˈnetɪk]	adj.	地磁的，有磁性的
mechanism	[ˈmek(ə)nɪz(ə)m]	n.	机械装置
merge	[mɜːdʒ]	vi.	合并，融合
mimic	[ˈmɪmɪk]	vt.	模仿，模拟
Non-Uniform Rational B-Spline (NURBS)			非均匀有理B样条
normals			法线，法向
obstruction	[əbˈstrʌkʃ(ə)n]	n.	障碍，阻碍
omit	[əˈmɪt]	vt.	忽略，省略
open-loop		adj.	开环的
operation	[ˌɑpəˈreʃən]	n.	操作
Operation Navigator			操作导航器
overhang	[əʊvəˈhæŋ]	n.	悬垂
pad	[pæd]	n.	衬垫
parent group			父节点组
Pass	[pɑːs]		刀路
peck-drilling		n.	啄钻
peripheral	[pəˈrɪfərəl]	adj.	外围的，外部的
permanent boundary			永久边界
perpendicular	[ˈpɜːpənˈdɪkjʊlə]	adj.	垂直的，正交的
pixel	[ˈpɪksl]	n.	像素
planar milling			平面铣

plunge milling			插铣
polygon	[ˈpɒlɪg(ə)n]	n.	多边形
postprocessing			后处理
precision	[prɪˈsɪʒ(ə)n]	n.	精度，精确
profile		n.	侧面，轮廓，轮廓切削
projection	[prəˈdʒəkʃən]	n.	投影，投射
quadrant	[ˈkwɒdrənt]	n.	象限，象限角
ramp	[ræmp]	vi.	蔓延，使有斜面
ream	[riːm]	vi.	铰孔
Reference Coordinate System (RCS)			参考坐标系
rest mill			剩余铣
retract	[rɪˈtrækt]	vi.	退刀
rough	[rʌf]	vi.	粗加工
scallop	[ˈskæləp]		残余高度
screw	[skruː]	n.	螺旋，螺丝钉
segment	[ˈsegm(ə)nt]	n.	部分，段
serial communication			串行通信
servomotor	[ˈsɜːvəʊˌməʊtə]	n.	伺服电动机，继动器
silhouette	[ˌsɪluˈet]	n.	轮廓
spike	[spaɪk]	n.	尖峰
spindle	[ˈspɪnd(ə)l]	n.	主轴
spot-facing		n.	锪孔，扩孔
stagger	[ˈstægə]	adj.	交错的，错开的
stator	[ˈsteɪtə]	n.	固定片，定子
steepness	[ˈstɪpnɪs]	n.	陡峭，陡坡
stepover	[ˈstɒpˈəʊvə]	n.	步距
stepper motor			步进电机
stock	[stɒk]	n.	余量
subregion	[ˈsʌbˌrɪdʒən]	n.	子区域
tachometer	[tæˈkɒmɪtə]	n.	转速计；流速计
tap	[tæp]	vi.	攻螺纹
taper	[ˈteɪpə]	n.	锥形物
temporary boundary			临时边界
thread	[θred]	n.	螺纹
through-hole		n.	通孔，金属化孔
tolerance	[ˈtɒlərəns]	n.	公差

torque	[tɔːk]	n.	转矩，[力]扭矩
transducer	[trænz'djuːsə]	n.	传感器，变换器
traverse	[trə'vɜːs]	vi.	横越
trochoidal	[trəʊ'kɔidəl]	adj.	摆线的，摆线切削
tusk	[tʌsk]	n.	尖头，尖形物
undercut	[ʌndə'kʌt]	vi.	底切
underneath	[ʌndə'niːθ]	prep.	在…下面
undulation	['ʌndjʊ'leɪʃən]	n.	波动；起伏
variable-axis surface contouring			可变轴表面轮廓铣
violation	[vaɪə'leɪʃn]	n.	妨碍
wire-cut	['waɪəkʌt]	adj.	线切割
Work Coordinate System (WCS)			工作坐标系
zig	[zɪg]	n.	单向切削
zig with Contour			单项轮廓切削
zig-zag	['zɪgzæg]	n.	之字形，往复式切削
Z-level milling			等高轮廓铣

Reference

[1] Michael F. Machining and CNC Technology[M]. New York: McGraw-Hill Higher Education, 2013.
[2] Mike M. CNC Programming Principles and Applications[M]. Beijing: China Machine Press, 2008.
[3] Robert Q. Computer Numerical Control[M]. New Jersey: Prentice Hall, 2004.